CONSTRUCTION QUALITY
and
QUALITY STANDARDS

OTHER TITLES FROM E & FN SPON

'A' Time
The busy manager's action plan for effective self management
J. Noon

Architectural Management
M. P. Nicholson

An Introduction to Building Procurement Systems
J. W. E. Masterman

Construction Economics in The Single European Market
B. Drake

Construction Conflict Management and Resolution
P. Fenn and R. Gameson

Estimating Checklist for Capital Projects
2nd Edition
Association of Cost Engineers

Just-in-Time Manufacturing
T. C. E. Cheng and S. Podolsky

The Management of Quality in Construction
J. L. Ashford

Integrated Construction Information
Edited by M. Betts and P. Brandon

Management Quality and Economics in Building
P. Brandon and A. Bezelga

Managing Construction Worldwide
(3 volume set)
Edited by P. Lansley and P. A. Harlow

The Multilingual Dictionary of Real Estate
L. van Bruegel, R. Williams and B. Wood

Post-construction Liability and Insurance
Edited by J. Knocke

Project Control of Engineering Contracts
M. O'C. Horgan and F. R. Roulston

Project Management Demystified
Today's tools and techniques
G. Reiss

Profitable Practice Management
For the construction professional
P. Barrett

Risk Analysis in Project Management
John Raftery

Spon's Budget Estimating Handbook
2nd edition
Tweeds

Spon's European Construction Costs Handbook
Davis, Langdon & Everest

Value Management in Design and Construction
S. Male and J. Kelly

Effective Speaking
Communicating in speech
C. Turk

Effective Writing
Improving scientific, technical and business communication
2nd edition
C. Turk and J. Kirkman

Getting into Print
P. Sprent

Good Style
Writing for science and technology
J. Kirkman

Journals

Construction Management and Economics
Editors: R. Bon and W. Hughes

Building Research and Information
Editor: A. Kirk

For more information on these and other titles please contact:

The Promotion Department
E & FN Spon, 2–6 Boundary Row, London, SE1 8HN.
Telephone 0171 865 0066.

CONSTRUCTION QUALITY
and
QUALITY STANDARDS

The European perspective

George Atkinson
Consultant Architect

E & FN SPON
An Imprint of Chapman & Hall

London · Glasgow · Weinheim · New York · Tokyo · Melbourne · Madras

Published by E & FN Spon, an imprint of Chapman & Hall,
2–6 Boundary Row, London SE1 8HN, UK

Chapman & Hall, 2–6 Boundary Row, London SE1 8HN, UK

Blackie Academic & Professional, Wester Cleddens Road, Bishopbriggs, Glasgow G64 2NZ, UK

Chapman & Hall GmbH, Pappelallee 3, 69469 Weinheim, Germany

Chapman & Hall USA, 115 Fifth Avenue, New York, NY 10003, USA

Chapman & Hall Japan, ITP-Japan, Kyowa Building, 3F, 2-2-1 Hirakawacho, Chiyoda-ku, Tokyo 102, Japan

Chapman & Hall Australia, 102 Dodds Street, South Melbourne, Victoria 3205, Australia

Chapman & Hall India, R. Seshadri, 32 Second Main Road, CIT East, Madras 600 035, India

First edition 1995

© George Atkinson

Typeset in 10½/12 Times Ten by Florencetype Ltd, Stoodleigh, Devon

Printed in Great Britain by Clays Ltd, St Ives plc

ISBN 0 419 18490 2

A catalogue record for this book is available from the British Library

∞ Printed on permanent acid-free text paper, manufactured in accordance with ANSI/NISO Z39.48-1992 and ANSI/NISO Z39.48-1984 (Permanence of Paper).

Contents

Foreword

Pierre Chemillier

Le Président, Conseil Général des Ponts et Chaussées

Achievement of construction quality is a central goal of practitioners throughout Europe. They are helped by many national, European and international bodies, ranging from the United Kingdom's Building Research Establishment and my own country's Centre Scientifique et Technique du Bâtiment (CSTB) to European and international organizations like the European Union of Agrément (UEAtc) and CIB.

In their move on the creation of a Single European Market, experts from governments, industry and the professions are coming together to pool experiences. In the Interpretative Documents, which are a bridge between performance of construction products and the requirements from building and civil engineering works for health, safety and some aspect of amenity and convenience or energy savings, the European Union's Standing Committee for Construction has, as George Atkinson suggests, provided the basis of a European 'house style'.

In *Construction Quality and Quality Standards*, George Atkinson looks at the different ways European member States approach quality. He draws lessons for cooperation as well as providing detailed guidance on national institutions and the sometimes complex machinery for European cooperation. This book will be very helpful for all people acting in the Construction sector at European level.

Introduction

This book is about quality in construction. It is also about codes and technical specifications,[1] which help – and sometimes hinder – practitioners in the achievement of quality. It describes their making within the national scene and the wider, European scene, and the many different organizations involved.

The book marks a path through the European Union forest of regulations, directives, harmonized standards, CE marks and other legislative tools. It is not an easy task. The path chosen is first to look at long-established native trees, particularly in the UK but also in France, Germany and other Member States, and then to look at those freshly planted by the Brussels foresters, particularly where they are intended to replace native varieties.

Before doing so, the meaning of quality in construction is discussed, and how it is achieved – or not achieved – in traditional practice. The book is addressed to the many different groups who have an interest in construction quality:

- owners and users of buildings and civil engineering works: householders and public and private tenants, especially those authorities whose purchasing arrangements for works and supplies of products and services are being changed through EU legislation;
- construction practitioners: architects and other designers, construction economists, builders, installers of services and specialist equipment, and producers and suppliers of materials and equipment, all with a double interest – to promote business, and, by reducing abortive and defective work, limit avoidable costs;
- public authorities, central and local, with statutory responsibilities for public health and safety, and possibly amenity and convenience, and conservation of national resources such as energy, fuel and water;
- standards-making bodies, certification and inspection bodies, testing laboratories, technical control offices, and construction research and information organizations, whose work is to develop and promote the use of the technical specifications through which construction quality is evaluated;
- financial groups with interests in construction: public agencies, commercial firms and private investors; banks, building societies and savings institutions; firms and syndicates of insurers covering fire

and other damage risks, or indemnifying designers, builders and others against claims for negligence under contract or tort.

Because there are many participants in the construction process, and because construction has features that differ from manufacturing industry, achievement of construction quality is not easy. Even among practitioners, various groups place different priorities on its satisfaction and may not agree on who was responsible when works fail to achieve quality.

Much of the book is about the various and varied initiatives that have resulted from the decision in May 1985 to create a Single European Market. Misleadingly, 1992 was much talked about as the year when these initiatives would reach fruition. We now know it will not be so, and the creative process will be ongoing for the rest of the 1990s and, probably, well into the 21st century.

The initiatives affect people as citizens, and as workers; and firms as businesses, and as employers. They affect the quality of the products made in, or imported into the EU, and the work done on those products. The impact of such initiatives on construction is being felt through EU legislation, such as 89/106/EEC, the Construction Products Directive,[2] and the many new European standards that will result; also national legislation that transposes this and other EU directives, like the UK Construction Products Regulations 1991, and regulations issued by the Health and Safety Commission on safety standards in workplaces. The initiatives include European legislation aimed at opening the procurement of public works and supplies to European competition.

These new initiatives have strengthened arrangements for standards-making at a European level. They have seen the setting up of new organizations such as the European Organizations for Technical Approvals, and for Testing and Certification.[3] They give a different status to national schemes for product approvals, like British Agrément Certificates, and for quality marks, like BSI Kitemarks. To bring them about, complex arrangements of assessment, testing and certification are being forged. For the British Standards Institution, European initiatives are seeing significant changes in the way standards for construction are handled, as drafting and updating responsibilities move from London to Brussels, although subject to major UK inputs in the work of standards-making.

In the short term, some of these initiatives will have little influence on construction quality. For some time, they may have little effect on the work of practitioners, whether they design or build, although their professional and trade organizations are much involved in ensuring that results benefit rather than disadvantage their members. They are of more immediate interest to manufacturers of construction products, and to certification, inspection and testing bodies, including firms who wish either to set up in one of the Member States, or to export products and equipment to the EU.

Suppliers of products and equipment established in more than one European Member State are already aware of many of the issues involved in creating a Single European Market, as increasingly as those in the European Economic Area. The initiatives will become of major importance for regulators as European standards and other technical specifications replace familiar national standards and specifications; and for purchasing authorities as they become mandatory under European procurement procedures.

Part One of this book looks at construction quality and its achievement, first in the meaning of building well and the crucial role of the client – taking, as examples, the commissioning of the design for four outstanding European buildings. It then looks at experience with achievement of long-term quality in a number of buildings, mainly from the 1930s, and at some of the problems of its achievement in social housing.

There follows a discussion of the role of standards in the achievement of faultless construction, and a description of the standards-making process in the UK. A chapter is devoted to quality assurance and the contribution of quality systems standards. A review of the meaning of fitness for use, and the contribution of the agrément concept, serves as an introduction to the various initiatives resulting from the Single European Market.

Three chapters are devoted to important items of European legislation: 89/106/EEC, the Construction Products Directive; the series of directives on works and supplies; and the European technical specifications – European standards and technical approvals – that support the legislation.

The ways in which construction quality is being achieved in six European countries, and the key organizations involved, are then described. The last two chapters are devoted to a discussion of a number of unresolved issues, mostly arising in the course of implementing Union legislation; and to a review of some of the lessons to be drawn from the many-sided task of achieving quality in construction.

Part Two is of a different character, being devoted to notes on the many bodies – international, European and national – concerned with construction and its quality. There is a reference section in which key data on Member States of the European Union and European Free Trade Association are summarized.

1 For a definition of 'technical specification' and other terms mentioned in the book, refer to Part Two, Chapter 16.
2 Throughout the book, this Directive will be referred to as 89/106/EEC.
3 Descriptions of these and other international and European organizations will be found in Part Two, Chapter 14.

Construction Quality and its Achievement

PART

1

Building well 1

In Architecture, as in all other Operative Arts, the End must direct
the Operation. The End is to Build well. Well Building hath three
Conditions; Commodity, Firmness, and Delight.

The Elements of Architecture collected by
Sir Henry Wotton Knt (1624)

Sir Henry Wotton drew for *Elements of Architecture* on contemporary
and earlier works in an extensive architectural library, formed when
he was English ambassador to the Venetian Republic during the early
years of the 17th century. He was two years old when Palladio's
Quattro Libri del' Architettura was published, a book that he owned
with the 1556 Barbaro translation of Vitruvius, and an almost contemporary treatise by Philbert de l'Orme, one of the first books to
discuss a key issue in quality – how a client may distinguish between
a wise and a foolish architect.

Wotton would have seen Palladio's Venetian churches and his villas
on the Venetian terra firma in their first youth, and his book includes
comments on Italian building traditions. Wotton's *Elements* are, therefore, an appropriate introduction to an attempt to give a European
perspective to construction quality. His much quoted 'The End is to
Build Well' sums up, in six words, the theme of this book.

1.1 MEANING OF QUALITY

Quality is one of the aims of standardization . . . the quality of a
product or a complete building or other construction is the totality of its attributes that enable it to perform a stated task or to
fulfil a given need satisfactorily for an acceptable period of time.

For building and civil engineering . . . a satisfactory product,
although essential in itself, is not on its own sufficient. It must be
incorporated in the design and construction in a correct manner.

In buildings, more defects and failures arise from inadequacies in
the treatment of products in design and construction than from
shortcomings in the products themselves.

BS PD 6501: Part 1: 1982.

BOX 1.1 QUALITY IN DESIGN, CONSTRUCTION AND USE

Quality of design process derives from:

- reliability of initial brief
- reliability of all information used as basis of the design, and selection of products
- reliability of design solution and detailed specification
- reliability of estimates of quantities of materials and labours required and their costs, of management and site overheads, and predictions of possible contingencies
- reliability of calculations relating costs to benefits, including tangibles like economies in energy consumption
- experience of designer in judging whether estimate of total cost of project is realistic, and will meet requirements of client

Quality of construction process derives from:

- reliability of organization, procedures and skills of builder to interpret the design, marshal required resources and provide the end product in accordance with design and specification, and at contracted price
- a workforce of appropriate skills
- products of specified quality

Quality of products derives from:

- reliability of all the materials, products, components and equipment supplied to the site, and their handling, storage and protection on site

Quality of building in use derives from:

- reliability of commissioning of installations and inspection of work on handing over, and making any corrections required
- reliability of the maintenance programme
- reliability of management of building in use, including assurance that any alteration to building or modification to installations will not impair performance or quality achieved

From classical times, writers like Vitruvius have attempted to define the meaning of quality, usually in fairly general terms. More recently, interest in construction quality has brought into more or less general use a range of definitions, some specific to the increasingly wide international and European use of the ISO 9000/EN 29000/BS 5750 series of standards for quality systems; others from concern with the quality of construction products.

The term 'quality' has different meanings for different people. There are subtleties in the word's use in the English language. It has been used to describe people, their skills and, in earlier centuries, rank in society; but also the quality of things, where two uses merit quotation. The first is appropriate to both buildings and construction products: 'a particular class, kind or grade of anything, as determined by its quality', dated 1656 in the *Shorter Oxford English Dictionary*. The second, quoted from George Bernard Shaw's *John Bull's Other Island*, is particularly apposite for quality assurance: 'There are only two qualities in the world: efficiency and inefficiency'.

The French use *qualité* as a term meaning 'property', as well as an indication of degree of excellence. It is left to the Germans to reserve *qualität* for a suggestion of grade, e.g. *erste qualität* (prime quality), and to use *eigenschaft* when referring to an attribute, characteristic or property.

The quality of a building derives not only from the quality of its design and of the process through which the design was developed, from the quality of the construction process and the care taken in translating the design into practical shape, and from the quality of products used and equipment installed, but also from the way it is used, and the quality of building management and maintenance.

In this part of the book, achievement of quality at the building level is discussed: first, against the findings of a BRE survey of quality and value in building; then, in looking at how quality was achieved in four major European buildings, and the interaction between commissioning of their designers and achievement; and, finally, against the issue of quality in traditional construction.

1.2 QUALITY AT THE BUILDING LEVEL

Although the preparation of the brief and the layout design are crucial, all subsequent parts of the design, construction and use of the building have to be done well to ensure quality and value.

> The achievement of quality and value is therefore intrinsic to the whole process and cannot be regarded . . . as an optional extra.
> *A survey of quality and value in building,*
> BRE Report (1978)

Concern with the achievement of quality at the building level is not new. In the last century, Alfred Bartholomew, architect and early editor of *The Builder*, prefaced his *Specifications for Practical Architecture* with an essay on 'the decline of excellence in the structure and in the science of modern English buildings; with proposals of remedies for those defects'. Despite the excellent materials that 'the English Architect has in modern times at his disposal, . . . the actual practical building of this country retrogrades sadly both in goodness and wisdom'.

If such was the situation in Victorian England when the traditional building crafts still flourished, how much more critical is the situation today, when avoidance of latent defects has become a first priority. Bartholomew's solution was 'the Foundation of a Great National College, for the Study and Regulation of Architecture throughout the British Dominions, for the Examination of Students and Professors of Architecture, and Artificers in Building, for the granting Honorary Degrees to Proficients therein of various stages of Maturity, and for the Conservation of Public Buildings'.

Solutions promoted today range from the joint training of the construction professions, and mandatory continuing professional development, to peer reviews and quality assurance not only of products, but of management systems of firms and performance of personnel during the whole construction process: briefing, design and construction. These matters, and the standards and other technical specifications needed for support, are central themes in this book.

1.3 BRE STUDY: *A SURVEY OF QUALITY AND VALUE IN BUILDING*

In the late 1970s, helped by several colleagues, M.E. Burt, then deputy director of the UK Building Research Establishment, set out a framework for establishing criteria for quality, performance and value. As far as was possible, he showed how these criteria could be evaluated against each other. Quality was defined as 'the totality of the attributes of a building which enables it to satisfy needs, including the way in which individual attributes are related, balanced and integrated in the whole building and its surroundings'. Its attributes are summarized in Box 1.2.

An important section was devoted to the role of the client, and his contribution in the initial decision-making process, starting with commissioning. The client might be an individual needing a house, or other modest building, who was, in the words of P. Peter writing on the French decennial guarantee, '*un profane* (layman) *en matière de construction*'. The client might be a substantial organization with a clear idea of the functional needs of the building commissioned, but with no capability in-house for design or construction. The client might be a public authority or what in French is termed *une personne morale*: a major property owner or developer, experienced in commissioning works, and having substantial professional resources for specifying requirements and ensuring their satisfaction.

Whichever his status and experience, the client alone has to make the first, and the most critical decisions affecting quality: what are his requirements, and in what way, and with what resources, are they to be satisfied? Excluding the decision to rent or buy works already constructed, he has a number of choices:

- appoint an independent practitioner to take his instructions and prepare a design, and, if the design appears to meet his needs, entrust him as the client's agent with its execution;

BOX 1.2 THE ATTRIBUTES OF QUALITY IN A BUILDING

The totality of the attributes of a building that enable it to satisfy needs, including the way in which individual attributes are related, balanced and integrated in the whole building and its surroundings.

Three groupings are defined:

- *external attributes* – relating to the effects of the site and its surroundings on the building, and the effects of the building on its surroundings;
- *performance attributes* – mainly related to the interior of the building which makes it operationally efficient and provides reasonable conditions for users (e.g. accommodation, environment, safety and security, use);
- *aesthetic and amenity* – terms used to describe both external and internal attributes of a standard higher than just needed to meet mandatory and performance requirements (e.g. external appearance, and internal appearance, and internal standards of comfort)

Some clients may require a degree of amenity, not only for its own sake, but because they feel that it may make some contribution to the competitiveness of their business, perhaps by improving public or staff relations.

Attributes of quality could alternatively be grouped under five heads:

- applicability – are they related to the needs of a single building or applicable to buildings?
- whose benefit – client and occupier, users, the community?
- whose decision – client, the community or State through legislation and regulations?
- are attributes quantifiable, difficult to quantify, or subjective?
- to what extent are they governed in time by the client's brief, the design, construction, or use?

- entrust design and supervision of execution to an in-house organization – existing or formed for the project; or
- contract with a commercial firm to provide a design, execute its construction, and possibly furnish services like arranging finance, and supplying equipment and furniture.

Each approach has its merits; each its drawbacks. For experienced clients, choice may not be difficult. But the inexperienced layman has no clear route to obtaining objective advice on the approach best suited to his needs. If, for example, he decides to entrust the work to an independent practitioner, or to an 'all-in' design–build contractor, he faces the problem of how best to choose the one likely to give good value and do a satisfactory job. Professional institutions, through their client advisory services, may offer impartial advice which, however, is likely to be restricted to practitioners who are members of that institution. To assess designs offered by different contractors against estimates of cost, and the contractor's capability of delivery of a building of specified quality on time, is not an easy task. That the firm has been quality assured, and is subject to surveillance, is a help; but it does not guarantee the quality of the product, only the quality of its management and that of the process of production.

1.4 IMPORTANCE OF THE DESIGN BRIEF

Each route to commissioning a building can, but may not, result in work that satisfies a client's needs. Whichever route is chosen, these needs have to be made explicit in a design brief. The form and degree of elaboration of briefmaking will vary. It will be influenced on the initial decision on commissioning, and whether the development of the brief, first into a design, and then into detailed specifications and drawings, are entrusted to an in-house team, to an independent practitioner as the client's agent, or a design–management contractor. Because quality of decisions at the briefing stage is crucial, where the client is a public authority or major property owner with a large building programme, he should see that procedures for briefing his own designers and outside consultants are included in the organization's quality management system.

The four European buildings described in Chapter 2 may, because of their public character and quality, be atypical, yet they present lessons in commissioning. Of special interest is the distinction drawn in the case of the Grand Louvre between the *maîtrise d'ouvrage* – the client organization, representing the French government through a special *Etablissement Public du Grand Louvre* within which there are services with financial, contractual and insurance responsibilities as well as personnel with responsibilities for programming, briefing and overseeing the works – and the *maîtrise d'oeuvre* – which included not only the chief architect I.M. Pei and his team, the official conservation architect for the Louvre, and a team of archaeologists but also a number of technical offices (*bureaux d'études techniques*) with responsibilities for detailed design work.

FURTHER READING

Definitions of quality and associated terms in English, French and German are given in Part Two, Chapter 16 of this book.

Sir Henry Wotton's *Elements of Architecture* were collected from

'the best Authors and Examples' while he was English Ambassador
to the Venetian Republic at the end of the sixteenth century. The first
theoretical work on the subject in English, his *Elements* were published
in 1624, two months after his return to England. The extract given
here is from a 1733 reprint in John Evelyn's *A Parallel of the Ancient
Architecture with the Modern*.

The *Elements*, like many early treatises, were based on the Ten
Books on architecture by Vitruvius, a Roman writing in the first
century BC, whose manuscripts were preserved, copied in the Middle
Ages, and printed first in Latin in 1486 and then in Italian. Sir Henry
would have known the Daniele Barbaro edition, published with
commentaries and illustrations by Palladio in Venice in 1562.

Among those who studied Vitruvius was the French architect,
Philibert de l'Orme, son of a Lyon master-mason, who spent some
years in Rome measuring and even excavating the antiquities. De
l'Orme's two books broke away from the Vitruvian model and dealt
also with practical issues. His *Nouvelles Inventions pour bien baster et
à petit frais* (1561) was on the construction of vaults and roofs. His
second book, *Le Premier Tome de l'Architecture* (1567), is of special
interest in that the first two sections deal with practical questions, such
as the relations of the architect to the client (called by de l'Orme *le
Seigneur*). The client must consider whether he can afford the works,
and whether its upkeep will ruin him. He must select his architect with
care but, having chosen him, must give him a free hand. The architect
and client must plan everything before they start building; but the
client must not change his plans as the building proceeds. In a final
section, de l'Orme, in two woodcuts, illustrates the wise and the
ignorant architect.

For an account of Sir Henry Wotton, see Eileen Harris, *British
Architectural Books and Writers 1556–1785* (Cambridge University
Press, 1990). There is a short account of de l'Orme in Anthony Blunt,
Art and Architecture in France, 1500–1750.

Two definitions of quality are to be found in Part Two, Chapter 16.
Both stress meeting user and client needs, faultless construction, and
striking a balance between a building and its surroundings. This last
aspect of construction quality was the central theme of a discussion
document, issued in September 1994 by the UK Department of the
Environment. In a preface, the Secretary of State wrote: 'Quality
affects us all'; also 'Quality is sustainable', and 'Quality pays. Good
quality is economic'. Most of the document was directed to the visual
environment in town and country. There is, however, a section on
'Better quality building' in which the remarks made four centuries ago
by the French architect de l'Orme on good briefing are made again.

The paper might have taken further the 1978 BRE report, *A survey
of quality and value in building*, discussed in detail in this chapter, and
the follow-up 1987 BRE report 95, *Better briefing means better
buildings*. That selection of the architect and design team is crucial to
building quality is discussed in Chapter 2. Pierre Maurin, secretary-

general of the association of French construction insurers, describes in a clear, logical manner the respective responsibilities, in France, of the client organization – *maîtrise d'ouvrage* – and of the project team – *maîtrise d'oeuvre*.

Holding an architectural competition is one approach to selecting an architect. Peter Collins, drawing on a period of study at Yale Law School, compares decision-making in architecture and the law in *Architectural Judgement* (Faber and Faber, 1971). There are also useful lessons to be drawn from case studies of architectural competitions for major public buildings. As well as the four buildings reviewed in Chapter 2, earlier case studies include Ian Toplis, *The Foreign Office: An Architectural History* (Mansell, 1987), and David Brownlee, *The Law Courts: The Architecture of George Edwin Street* (MIT Press, London and Cambridge, MA, 1984).

The Royal Fine Arts Commission in 1992 commissioned Judy Hillman to trace, in a short study, examples where clients of public buildings had failed to brief architects effectively (Judy Hillman, *Medicis and the Millennium: Government Patronage and Architecture*).

The role of the client 2

The first and perhaps the most critical decision affecting quality belongs to the client alone.

A survey of quality and value in building,
BRE Report (1978)

Four recent European buildings serve as illustrations of how commissioning and briefing are associated, and how, if the choice of designer and brief is handled well, architecture of quality is achieved. The buildings chosen, all designed and built between the late 1970s and the early 1990s, are: in Federal Germany, the Neue Staatsgalerie, Stuttgart, and the Museum für Kunsthandwerk, Frankfurt am Main; in France, the Grand Louvre project, Paris; and, in London, the National Gallery Sainsbury Wing. As public art galleries and museums, their requirements are broadly similar, although, as Table 2.1 shows, they differ in size and other features. There were also national differences in procedures, due in part to funding and client organization, and in part to the different approaches of their designers – one British and three American.

Three of the four architects – James Stirling and Michael Wilford (Stuttgart), Richard Meier & Partners (Frankfurt), and Robert Venturi and Scott Brown (London) – were commissioned as a result of architectural competitions. I.M. Pei (Paris) was chosen after an extensive review of the works of architects of known reputation by a delegate of the client, the French Ministry of Culture.

2.1 THE STUTTGART GALLERY AND FRANKFURT MUSEUM

Both architects of the two German buildings – the new wing of the Staatsgalerie, Stuttgart, and the Museum für Kunsthandwerk, Frankfurt am Main – were chosen as winners of architectural competitions. The architects of the Stuttgart building, James Stirling, Michael Wilford and Associates, London, were winners in an open international competition; the architects of the Frankfurt building, Richard Meier & Partners, New York, were winners of a competition limited to seven participants.

Table 2.1 Four European quality buildings

Project (Construction date)	Client (Bauherr/Maîtrise d'Ouvrage)	Architect (How selected)	Area (m²) (Cost: date)
Neue Staatsgalerie, Stuttgart (1977–1984)	Baden-Württemberg Land Hochbauamt 1 Stuttgart	James Stirling, Michael Wilford (International architectural competition)	15 300 (DM 89M: 1981)
Museum für Kunsthandwerk, Frankfurt am Main (1979–1985)	Stadt Frankfurt am Main, Amt für Wissenschaft und Kunst Hochbauamt Frankfurt am Main	Richard Meier & Partners (Limited architectural competition)	9925 (DM 43½M: 1981)
Grand Louvre, Paris (1983–1996)	Musée du Grand Louvre (Établissement public du Grand Louvre)	I. M. Pei & US and French associates (Personal client choice, after wide review)	160 200 (Fr 1997 bn: 1983–1989 Fr 3109 bn: 1989–1996)
Sainsbury Wing, National Gallery, London (1985–1991)	Trustees of the National Gallery (Client project management under NG Services Ltd) (Donors: Sainsbury family)	Robert Venturi, Scott Brown (Comprehensive review by expert selection team, followed by design interviews limited to seven selected architects)	111 556 (Not available)

In Federal Germany, considerable stress is placed on assurance of technical quality. Stringent production control requirements in the Länder (state) building ordinances, supported by a comprehensive system of technical standards, mean that virtually all building products undergo some form of technical assessment, usually by a third party. Structural work is checked by an independent licensed structural engineer (*Prüfingenieur*).

Unlike the UK, where apprenticeship is no longer the norm, Germany does not lack skilled craftsmen. As a comparative study undertaken for the Anglo-German Foundation showed, there still exists in Germany a well-organized system of small *handwerk* firms whose scope of work is defined by law. These firms are still able to generate a high level of skills through the statutory backing given to trade definitions and training standards, the obligation on school leavers to undertake vocational training, and the relatively low levels of training allowances agreed in collective bargaining.

The right to set up a business is obtained through proficiency tests. Similarly, in larger firms, there are differences at the foreman level, with the Germans insisting on a much higher level of technical and supervisory skills than is general in the UK. Work is often entrusted to trade contractors, directly employed by the client, rather than as subcontractors to a general contractor, who may have a largely managerial role and employ few building tradesmen.

The clients (*Bauherren*) – Land Baden-Württemberg, and Stadt Frankfurt am Main – both of the Stuttgart Gallery, and of the Frankfurt Museum, were responsible for project management through their respective official building offices (*Hochbauamt*), day-to-day control (*Projektleitung*) being in the hands of experienced building engineers. However, the composition of the design and construction teams differed, in part due to the fact that much of the design work for the Frankfurt Museum was undertaken in New York. Each team of architects was backed by strong specialist teams – fifteen different teams at Frankfurt, and seven at Stuttgart – and, as usual in Germany, there was statutory checking of structural work by independent licensed engineers.

At Stuttgart, work was largely undertaken by trade contractors employed directly by the client. At Frankfurt, the City of Frankfurt building department (*Hochbauamt*) had overall responsibility as client representative, but because much of the design work was undertaken in New York, a Frankfurt engineering firm – Frankfurter Aufbau AG – was responsible for day-to-day construction and financial management, the structure being reinforced concrete again under the control of an independent *Prüfingenieur* consultant. Specialist German consultants advised on foundations, thermal and acoustic design and building services.

The Stuttgart Gallery is clad with alternating bands of yellow travertine from Cannstat near Stuttgart and red sandstone from Weiler near Heidelberg, the plinth being shelly limestone (*Muschelkalk*) from

Bavaria; and, to coordinate masonry work, the three quarries were required to cooperate in the employment of a specialist building engineer, every single slab and its fixings being detailed as a 'rain-screen' with open joints held away from the structure by stainless steel fixings. The Frankfurt Museum is clad with vitreous enamel panels, the design of which was in the hands of a specialist Institut für Fassadentechnik, Frankfurt.

2.2 THE GRAND LOUVRE, PARIS

France has a long tradition of personal patronage of large public works by heads of state, in part for the glory of France but also for their personal prestige, the commissioning of works being entrusted to a senior administrator like Louis XIV's great public servant, Jean Baptiste Colbert, or, in the 19th century, Napoleon III's Prefect of the Seine Département, Baron Haussmann. The works may be controversial; but they are rarely technical or architectural failures. Also, France has a cadre of public works engineers, with a long tradition of service to the state, and experience in setting up client organizations able to manage large projects.

For the Grand Louvre, as with all large French public buildings, a special authority – l'Établissement Public du Grand Louvre – was set up as *maîtrise d'ouvrage* (client body). The museum director and seven conservator heads of departments who together formed a 'college', with the museum's general administration, acted as 'user body'.

Once the client organization was in place, the most critical issue was the choice of architect. In the case of the Grand Louvre, the French President, who took a personal interest in the project, entrusted the task to Emile Biasini, a senior official in the Ministry of Cultural Affairs. Having decided that the Grand Louvre project was too large and complex to be the subject of an architectural competition, M. Biasini started by meeting museum directors in Europe and North America, nearly all of whom mentioned the name of I.M. Pei, architect of the East Wing of the National Gallery, Washington, which had been completed a few years earlier. Crossing the Atlantic, Biasini inspected the East Wing, met Pei and recommended his appointment as architect.

When President Mitterand met I.M. Pei to confirm the appointment, he gave him a copy of *Le Journal de voyage du cavalier Bernini en France*, which tells how two centuries earlier another foreign architect, the Italian Gian Lorenzo Bernini, clashed with the French architectural establishment. Pei must have taken the hint, as the story of cooperation of the Pei office with French collaborators has proved to be very different.

Around Pei and his staff, a design organization – *maîtrise d'oeuvre* – was built up mirroring that of the client. It included official architects like Guy Nicot, *architecte en chef du palais du Louvre*, French and American associates, and technical design offices (*bureaux d'études techniques*) with responsibilities for sections of the work, and a team of archaeologists. Two of France's leading licensed technical control offices, Socotec and Veritas, provided independent checking of structural work and fire safety systems.

The Grand Louvre scheme is much greater in size, has taken longer to complete, and involves the restoration and a major reorganization of one of Europe's best-known buildings. Yet in quality of design and execution of works it is the equal of the two German and the London buildings.

2.3 THE SAINSBURY WING, NATIONAL GALLERY

Unlike the other three projects, the Sainsbury Wing was a 'private venture'. Until the Sainsbury family funded the work, the story of the National Gallery extension on the Hampton site had been one of government indecision and architectural failure. There were many reasons, a main one being the lack of a decisive *maîtrise d'ouvrage* body – only a triad of conflicting interests: an action-oriented Secretary of State for the Environment; a Treasury whose task was to avoid public funding; and Trustees uncertain how to provide an acceptable building in which the Gallery's fine collection of early Renaissance paintings could be shown with little or no chance of government funding.

Things changed following the Sainsbury offer. The Gallery needs were explicit, and the first task was to select an architect. As Emile Biasini had done for the Louvre project, the selection committee, representing the benefactors, the gallery trustees and director, and an architectural adviser, visited galleries and museums in Europe and the USA to exchange experiences with their directors and staff. A shortlist of six architects were asked to submit design proposals to a three-day meeting with the selection committee at the National Gallery, during which the six outlined their approaches to the design of the extension in relation to the main Gallery building and its surroundings. At the end of a series of challenging discussions, Robert Venturi, Scott Brown of Philadelphia was selected unanimously. But before a public announcement was made, several members of the selection committee made a second visit to America to meet some of Venturi's previous clients and see his office.

The Gallery director and senior staff, as clients, defined in as much detail as possible the kind of galleries required, and how the building as a whole should function. To accomplish its construction a charitable company of limited liability, the Hampton Site Company was formed to receive the funds, develop the site, and manage the project. It delegated its powers to a client steering committee chaired by Simon Sainsbury and consisting of donors, Trustees and members of the Gallery staff, which became effectively *maîtrise d'ouvrage*. Day-to-day control of work was entrusted to a subsidiary company, NG Services Limited. An associate British architect, who was appointed to advise on UK practice and regulations, carry out design coordination and inspection functions on the site, became, with Venturi and a number of specialist consultants, the *maîtrise d'oeuvre*. A construction management contractor was appointed to coordinate trade contractors. On completion, the building was handed over to the National Gallery Trustees.

2.4 THE IMPORTANCE OF TOTAL QUALITY MANAGEMENT

The two German buildings, and the Sainsbury Wing, are examples of largely traditional approaches to the achievement of quality in building. German and British practice differ in a number of ways, such as:

- manner of funding – State and City, in the case of the German buildings; private benefactors, in the case of the National Gallery – which was reflected in the client organization;
- a mandatory requirement by the German building control authorities for structural designs to be checked by an independent licensed engineer;
- the stringent requirements in Germany for quality assurance of products and equipment used, imposed under building ordinances; and
- the German tradition of craft apprenticeship and employment of *handwerk* firms.

Yet there were many common features, such as the procedures for selecting the architect, the way the client managed the projects, and the quality of the finished work.

Wider adoption of quality assurance procedures by the UK construction industry and professions should provide a framework through which an acceptable level of quality can be achieved. But it is not enough in itself. It would not, for example, overcome shortages of skilled supervisors and craftsmen, and the failure of government with employers' and workers' representatives to devise and implement a modern alternative to the now eroded traditional apprenticeship.

As the earlier history of abortive proposals for the National Gallery Hampton site and the long-drawn-out story of the British Library demonstrate, there are other factors that total quality management alone will not solve, such as the indifferent performance of clients, particularly central government, in commissioning and briefing design teams, and a failure, in the public service, to maintain an experienced professional cadre, similar to that in central government in France, and at *Länder* and city levels in Federal Germany, able to serve, when called on, either as members of the *maîtrise d'ouvrage* organization or, with private consultants and specialists, as part of a *maîtrise d'oeuvre* project team.

It is not possible here to explore the merits and drawbacks of architectural competitions in commissioning architects, which, as Ian Toplis tells in *The Foreign Office: An Architectural History*, are a perennial source of controversy. In *Architectural Judgement*, a book in which Peter Collins analysed judgements about architecture by comparing them with concepts evolved in the legal process, the merits and demerits of architectural competitions are examined. Much of the debate has been centred on the question as to whether the purpose of a competition is to choose an ideal design or an ideal designer. If it is the design, there is usually an outcry if an inexperienced winner proves unable to manage its execution. If, as in the four cases discussed here, it is to choose the ideal architect, then a need to ensure quality of design and execution points either to a limited competition or to careful investigations of past work, reputation with previous clients, and practice management of a few qualified firms. Collins summed it up in the following words:

> It would seem indisputable that competitions restricted to half a dozen nominated competitors are, from the judicial point of view, the most likely to produce reliable results.

FURTHER READING

Neue Staatsgalerie, Stuttgart, and the other three buildings, are described in a number of architectural periodicals. James Stirling has written on the competition and his design concept in James Stirling, *Die neue Staatsgalerie, Stuttgart* (Verlag Waltraud Krase, Stuttgart, 1984). There is a detailed description of the building in this book; also in *Neue Staatsgalerie, Stuttgart*, issued, also in 1984, by Staatliches Hochbauamt 1, Stuttgart.

An illustrated booklet, *Museum für Kunsthandwerk, Frankfurt am Main*, issued by Der Magistrat der Stadt Frankfurt am Main in 1985, describes this building, giving technical details of the project organization. Markus Lupertz describes both the Stuttgart Gallery and the Museum für Kunsthandwerk, Frankfurt a.M. in *New Museum Buildings in the Federal Republic of Germany*, published in 1985 in connection with an exhibition of recent museum buildings at the Deutsches Architekturmuseum, Frankfurt a.M. sponsored by the Goethe Institut.

'Le Grand Louvre' is one of nine major projects described in *Architectures Capitales: Paris 1979–1989* (Electa Moniteur, 1989); a description of the conversion of the north wing of the Louvre, previously occupied by the Ministry of Finance, to form l'aile Richelieu, is described in a special issue of *Connaissance des Arts*, *L'Avenir du Louvre* (1980). During construction work on the Grand Louvre, the Établissement Public de Grand Louvre published

Chroniques Grand Louvre: Journal d'Information de l'EPGL (1991–1992). Anthony Blunt, in *Art and Architecture in France 1500–1750* (Penguin Books, 1953), describes the rejection in 1667 of Bernini's design for the Louvre. For further information, contact EPGL Établissement du Grand Louvre, 153 rue Saint-Honoré, 75001 Paris.

Colin Amery tells the story of the National Gallery Sainsbury Wing in *A Celebration of Art and Architecture* (National Gallery, London, 1991).

Quality in traditional construction $\boxed{3}$

Every building of any importance is a solution of the problem of how to enclose within a visually pleasing and appropriate shell the space wherein human activities of one sort or another may be carried out with efficiency, in comfort and with dignity . . . The success of a building will depend on the vision of those who brief the designer, on the skill of the designer, and the good work of the builder, on the suitability of the materials in which the proposal takes material form and, by no means least, on the amount of money made available for the work.

All materials decay in time; the rate of decay can be retarded by good, or accelerated by bad detailing and workmanship . . . The weaknesses of well-tried materials are known, and allowed for in good design; but even the best designer may have to use materials with known shortcomings because, for a particular purpose, there exists at the time no economically acceptable alternative.

Introduction: *A qualitative study of some buildings in the London area* (1964)

3.1 ACHIEVEMENT OF LONG-TERM QUALITY IN BUILDINGS, MAINLY FROM THE 1930s

The case studies in Chapter 2 are directed at buildings completed in the very recent past, or in the case of the Grand Louvre still being completed. Authoritative studies of how well quality has been maintained over longer periods of time are scarce. A study undertaken by the Building Research Station in the 1960s is one of the few made. The Station knew much about problems resulting from poor technical performance of structures and materials. To present a more balanced approach to building practice, it investigated the overall performance over time of a number of buildings erected by reputable builders, who employed craftsmen capable of carrying out work of the quality required by architects of experience. The aim was to discover how far traditional practices achieved buildings of quality.

Thirteen London buildings were selected, all but one built this century and most during the inter-war period. They included the professional headquarters of the Institution of Civil Engineers, Great

George Street, and of the Royal Institute of British Architects in Portland Place. Most were public or institutional buildings; only two were commercial buildings.

A team of three – two architects and a structural engineer – carried out inspections over a period of years, their findings being reported in National Building Studies Special Report 33: *A qualitative study of some buildings in the London area.* One member of the team continued the study with inspections of two pre-war buildings: the De La Warr Pavilion, Bexhill-on-Sea (architects Dr Erich Mendelsohn and Mr Serge Chermayeff), and the Gilbey Building, London NW1 (architect Mr Serge Chermayeff). His findings are reported in National Building Studies Special Report 39.

Each building was studied as a functioning entity, with a view to discovering how far its designers' intentions and expectations had been satisfied, and of assessing the actual efficiency of the structure and materials in relation to each other, and to the general planning of the building. The investigators' concern was with the effect of materials and techniques as a whole on the quality achieved – a concept

> intrinsically not a subject for measurement, and difficult to define except in terms of durability, serviceableness, good weathering properties which enhance appearance, and freedom from heavy maintenance.

Except for the Bexhill Pavilion and the Peter Jones Store, Sloane Square, London, the buildings studied were good examples of traditional practices, and the overall conclusion of the investigators was:

> The buildings studied had continued to serve their intended purpose with few disabilities – the more creditably in view of the wartime hazards to which they were subjected. The recorded imperfections in materials in exterior or interior exposure, though numerous, were mostly of a minor character, impairing the appearance rather than the structural or functional efficiency of the buildings.

Other findings are grouped under three heads:

- changes in specification, keeping of records and uses of Codes of Practice and British Standards;
- the place of innovations;
- the different timescales of a building's structure and its services.

3.2 CHANGES IN SPECIFICATION DURING CONSTRUCTION

The proper recording of changes in specifications during the process of construction is a key element in any quality management system. It is therefore noteworthy that the BRS investigators found, in almost all cases studied, that the architects' specifications had never been followed in their entirety, that copies of variation orders had in no instance been preserved, and that changes of intention could be inferred only from visual evidence in the building; . . . seldom do accurate records appear to have been kept of alterations made in the course of the contract or subsequently, whether to the plan, to the constructional details or to the services.

Furthermore, in most specification records, there was a lack of technical information

due to the inclusion of prime-cost items for work to be undertaken by subcontractors, whose records of the job are no longer available; indeed many of the firms no longer exist.

For buildings erected before 1940, references to British Standards were infrequent or non-existent:

Many specification clauses were couched in a traditional form inherited from previous generations and which changed but slowly between 1910 and 1939. It was apparent that in a number of cases such clauses could not have been followed literally.

3.3 THE PLACE OF INNOVATION

The buildings studied were designed before the introduction of Agrément, and, except for the Peter Jones store and Bexhill Pavilion, innovations were largely restricted to building services. In the University of London Senate House, the architect went further in deliberately designing for very long life. Except for the Library Tower, steel framing was rejected as a main structural system 'on the grounds that there was less than seventy years of experience from which its probable endurance could be assessed'. The structure was of load-bearing brickwork – three qualities being specified, piles were designed to take their load – even if the reinforcement should rust away, and, except for brickwork up to first-floor level, lime mortar, carefully prepared on the site, was specified in preference to a lime–cement mortar, the long-term effect of which on the limestone masonry the architect considered to be suspect'. Yet, despite such precautions the team found that

cracking and spalling of the edges of the concrete roof-lanterns show how difficult it always has been, even in the 1930s when workmanship of the highest quality was readily available, to get reinforcement properly placed to ensure sufficient cover.

In marked contrast to the other buildings studied, the Bexhill Pavilion was non-traditional both in its design and in its construction. It was won in open competition by Dr Erich Mendelsohn, who had recently come from Germany to join for a time Serge Chermayeff in a London partnership. The architects' original intention was to build the Pavilion entirely in reinforced concrete. But it happened that the structural engineer, Felix Samuely, had also recently arrived from Germany, where he had worked on Berlin's first all-welded steel building. Faced with the architects' need to cut costs, Samuely proposed as an alternative a welded frame, the external walls being built in thin mesh-reinforced *in situ* concrete.

In adopting this and other innovations, little attention was paid to the risk of severe corrosion. What has happened over the years gives an invaluable insight into the kind of quality problems that can result from innovation on a shoestring. The danger of corrosion in a seafront exposure was known. Aluminium alloys were considered but, being in early stages of development, were rejected. It was difficult to obtain galvanized window frames of the large size required; and the final decision to use painted steel was probably taken on the grounds of initial economy even though, in the event, initial savings in capital costs must have been exceeded many times over in the course of a losing battle against corrosion.

The Peter Jones store, by contrast, showed that innovation need not result in costly maintenance. The facade to Kings Road and Sloane Square was an early example of a glass and metal curtain wall, the first length having stainless steel, and the second bronze mullions. Detailed examination by the BRS team showed that the curtain wall construction was still in good condition:

> To the achievement of this end, careful design, appropriate choice of materials with adequate protective treatment and efficient maintenance had all contributed. The facade is unusual, and was much more so at the time it was designed; it was indeed the outcome of careful thought and experiment, and its well-kept appearance is now regarded by the owners as a remunerative form of publicity not to be jeopardized by parsimony.

3.4 DIFFERENT TIMESCALES OF STRUCTURE AND SERVICES

The effect of rapid changes in building services technologies was another lesson drawn from the study. The life of a building service is often significantly shorter than that of the structure in which it is installed, a matter demonstrated in the more traditional buildings investigated as well as in the more innovative. Embedded heating pipes gave trouble. In the case of the Peter Jones store, large areas of glazing made it difficult to keep the building adequately warm with the original heating system. With the technology then available it was difficult to adjust internal temperatures to keep pace with very quick external temperatures changes.

Changes in the technology of lighting meant that original installations were soon outdated, an example being that in the RIBA building, where the original lighting design depended on a large number of concealed lights, which gave a bright ceiling in comparison with which everything else was in the dark.

Further changes to the building services will certainly have been made in all the buildings that the team investigated, a fact that makes even more relevant to building management the team's comments on the absence or loss of records of alterations made either in the course of a contract or subsequently:

> This absence of reliable records of a building as completed or altered not only makes the task of assessing performance more difficult than it need otherwise be; it also complicates further alterations and repairs and may increase their cost. At the very least, records of original service runs should always be kept in a safe place and careful notes made of any subsequent modifications. Omission to do so frequently causes trouble and expense when the time comes to renew them or the materials in or behind which they run.

The BRS team's view on codes of practice in design, made in a general commentary at the end of the main study, merits recording here:

> The Codes of Practice in present use are designed to obviate most of the shortcomings of the past and in time they will no doubt bring about a closer approximation between specifications as drafted and buildings as completed. No building of character could, however, be produced merely by following books of rules, diagrams or the recommendations of committees. Such aids may be of practical use amid the complexities of present-day design and for the prevention of inferior work, but it should be remembered that the most inspired and durable works of architecture in the past were not created by such tools, but by intuition combined with keen observance of past performance and sound common sense on the part of the architect himself. To these fundamental attributes of good building there are no short cuts, and there are no substitutes for them.

3.5 ACHIEVEMENT OF QUALITY IN SOCIAL HOUSING: A BRE INVESTIGATION

> Why it is better to 'get it right' than 'put it right' . . . For a proportion of the faults, the cost of the fault-free detail is no more than that of the faulty original; for others the cost is a little higher . . .

> The most striking result (of a cost analysis) is that, on the average, the cost of a faulty design detail plus its consequent remedy was about three times the cost of a fault-free detail.
> R.B. Bonshor and H.W. Harrison, BRE Report (1982)

The achievement of quality in low-cost housing is less easy than in the buildings discussed in the previous sections. Its achievement requires at least as good management organization and skills, and care by designers and builders. Much has been written about defects in social housing, although the information is too often anecdotal, and relates to small samples in the total post-war housing stock. However. a second investigation by BRE, this time directed at traditional housing under construction, gives useful information on how far quality can be achieved.

Rather than wait some years before latent defects show up as visual damage, houses during construction were studied. The aim was to identify faults in design or execution – defined as departures from good practice, the criteria being based on requirements of Building Regulations, British Standards and Codes, and published recommendations of recognized authoritative bodies such as BRE. Departures from drawings and specifications were also regarded as faults – unless the documents themselves were faulty. Against these criteria, the investigators identified just over 120 types of fault.

The study was limited to low-rise housing, 1725 dwellings on 15 sites being inspected. Just over half the types of fault observed were in the external envelope – walls, roof, windows and doors – with one-fifth of these being seen during construction of external walls, and another one-fifth in that of the roof. An eighth of all faults were observed in connection with windows and doors; and a further tenth in floors. The remaining fault types were related to the building services, external works, substructure, separating walls and stairs (Table 3.1)

In low-rise housing, most work on site is directed to the construction of roof, walls and floors; and, as expected, a large proportion of site faults were found among items belonging to bricklayer and carpenter–joiner trades. The cause of three-quarters was thought to be lack of care, such as damage in handling or wrong positioning of manufactured components. Just under a fifth were put down to a designer's placing undue reliance on site skills to solve difficult

Table 3.1 Traditional housebuilding: attribution of faults by element of construction (from BRE survey of housing, 1982)

Element	percentage of 955 recorded faults	
External walls	20	———————————
Roofs	19	———————————
Doors/windows	13	—————————
Floors	11	————————
Services	9	——————
Planning, layout, external works	8	—————
Substructure	7	————
Internal partitions		
Separating walls		
Stairs: each element	4	——

constructional details. Comparatively few faults were judged to be due to a tradesman's lack of craft knowledge, or manual dexterity. Most fault types were judged to have arisen from work on site; a comparatively small number from manufactured products; and the balance from faulty designs, or poor communication of design intentions to site personnel.

Lessons drawn from the study have been used in two BRE information series directed at designers and builders. From the two series, Defect Action Sheets and Good Building Guides, four conclusions stand out:

- A quarter of all fault types of a design origin were departures from either Building Regulations or British Standards cited by Building Regulations.
- Few designers had ready access to the complete range of authoritative advice contained in British Standards and Codes of Practice.
- Although examples of poor products were observed, they were less of a problem than misuse of products of acceptable quality, including poor site storage and handling.

An experienced team of investigators can identify potential faults in design and work under construction; but it is a matter of chance whether or not a fault shows up as a building defect for years, if ever. An exceptional wind storm may cause a faulty roof to collapse when otherwise the fault would remain undiscovered. A change in occupation may, for example, greatly increase persistent condensation, the resulting mould growth showing up cold bridges not apparent during previous occupation. Structural cracks, a result of shallow foundations in a clay soil, may only appear during a severe drought.

3.6 BRE INVESTIGATIONS INTO NON-TRADITIONAL HOUSING

From the late 1980s to 1990–1991, BRE has published a series of reports on the performance of many systems of non-traditional construction, the results of surveys of post-war housing, mostly owned by local authorities. Methods of construction included steel-framed and steel-clad systems, concrete and reinforced concrete systems, including large-panel systems and no-fines concrete. Positions where there was deterioration were identified to help subsequent inspections by owners' surveyors. Based on the information collected, guidance was given on remedial work.

Among non-traditional housing built in the post-war period, the largest number were flats, maisonettes and houses constructed with walls of either load-bearing no-fines concrete or framed structures with no-fines concrete walls. Not regarded by many as an industrialized system, they did in fact make a major contribution to post-war housing programmes, especially in Scotland; and the BRE studies provide a useful addition to our knowledge of their long-term performance.

The studies differ from the traditional housing study reported earlier in that the housing had been completed and occupied for many years. Some houses, still in local authority ownership, were being upgraded; others had been sold to occupiers, or handed over to housing associations. Much of the technical feedback on performance came from local housing authorities, and showed that:

• there were few significant structural problems originating in no-fines construction, with any major problems arising from local ground conditions rather than the structural design;
• by modern standards, much of the earlier housing was poorly insulated, hard to heat and prone to condensation, largely a result of the standards set by housing ministries at the time;
• owing to faults in design and construction, there were cases of rainwater penetration and deterioration of external rendering; and
• there were many examples of wooden windows and door frames that had rotted, a common problem in post-war housing due to poor designs and specifications.

The investigations into the longer-term performance of no-fines and other non-traditional forms of construction confirmed the previous findings of the traditional-housing study that the average cost of putting a fault right after construction could be at least five times the extra cost of getting it right in the first place. This was particularly true in the case of no-fines concrete, where the initial installation of adequate thermal insulation, and of windows and doors of acceptable quality, would have avoided much more costly repairs later, as well as the management costs of housing authorities dealing with tenants' condensation and dampness problems. It has been costly for local authorities, and the national economy, that successive governments ignored the 1944 Dudley Committee report on the design of dwellings on 'standards of construction':

• Durability:

> In their capacity as landlords, local authorities are concerned with durability in relation to maintenance. During the 'economy drive' of 1931 there was, we believe, a tendency on the part of the Ministry of Health (then the responsible housing ministry in England and Wales) to force local authorities to accept a lower standard of specification and materials than they themselves desired. There may be a temptation to revert to this policy . . . we are convinced that to do so would be a great mistake.

• Thermal insulation:

> We have been impressed by the need for greater economy in fuel. One means of achieving this is by improving the thermal insulation of the structure of new houses.

An examination into the roles of the different authorities – central government ministries, local housing authorities, professional and industry organizations – that set, or fail to set, appropriate quality standards, would require lengthy and much deeper examination than is possible here. They do, nevertheless, add weight to two matters already discussed in the more general contact of building quality.

The first is the need to identify and make explicit, in the process of 'well building', the roles and responsibilities of the two components in its achievement, best expressed in the French collective terms: the *maîtrise d'ouvrage*, or client team, and the *maîtrise d'oeuvre*, or construction team. The first, in a social housing programme, must include both central agencies with responsibilities for setting quality standards as well as funding policies and local agencies with responsibilities for commissioning and ensuring that quality standards are achieved. The second involves all the constructors – designers and builders – whether employed directly or not by the commissioning authority.

The second is the need to consider any major project in terms of total quality management whether in the social–political or business–commercial areas of construction.

FURTHER READING

The Building Research Station qualitative studies of a number of buildings, mainly built in the 1930s or earlier, are reported in S.B. Hamilton, H. Bagenal and R.B. White, *A qualitative study of some buildings in the London area*, National Building Studies Special Report 33 (HMSO, London, 1964), and R.B. White, *Qualitative studies of buildings*, National Building Studies Special Report 39 (HMSO, London, 1966). The first report covers: the Royal Horticultural Society's Second Hall, Westminster (1928); the Methodist Missionary Society's Headquarters, London NW1 (1939–1945); Peter Jones Store, Sloane Square, London SW1 (1932–1937); the Town Hall, Watford (1939); 24 Chiswell Street, London EC1 (1954); the RIBA, 66 Portland Place, London W1 (1934); Duveen Extension, National Portrait Gallery, London WC2 (1933); King Edward VII Galleries, British Museum, London WC1 (1907–1913); the Town Hall, Greenwich; the Institution of Civil Engineers, London SW1; the Senate House, University of London (1936); Seymour Hall, Marylebone, London W1 (1937); and Queen Anne's Mansions, London SW1 (1873–1889). The second report covers the De La Warr Pavilion, Bexhill-on-Sea (1934–1935) and the Gilbey Building, London NW1 (1936).

There have been a number of BRE reports on post-war housing, among the more comprehensive being R.B. Bonshor and H.W. Harrison, *Quality in traditional housing. Vol. 1: An investigation into faults and their avoidance*, BRE Report SO32 (HMSO, London, 1982). Volume 2 is an aid to design, based on the results of the investigation; Volume 3 is an aid to site investigation.

Among other BRE reports is B.R. Reeves and G.R. Martin, *The structural condition of Wimpey no-fines low-rise dwellings*, BRE Report BR 153 (BRE Publications, Watford, 1989), and a summary report of the series *The structural condition of prefabricated reinforced concrete houses, designed before 1960*, BRE Information paper 10/84 (1984). A full list of the series is to be found in the *BRE Bookshop Catalogue*.

The background to prefabrication, non-traditional and industrialized system building is given in R.B. White, *Prefabrication: A history of its development in Great Britain*, National Building Studies Special Report 36 (HMSO, London, 1965); and Barry Russell, *Building Systems: Industrialization and Architecture* (John Wiley, London, 1981).

The route to faultless building

4

Quality should be considered in a relative sense, i.e. in relation to fitness for stated purposes, rather than in an absolute sense.

BSI PD 6501: Part 1: 1982.

4.1 ACHIEVEMENT OF FAULTLESS BUILDING

Achievement of faultless building is the aim of assured quality in construction. It cannot be achieved unless the design, and resulting construction works, follow the acknowledged rules of technology. Many factors influence the achievement of quality, ranging from the reliability of the initial brief, and the choice of the design team, to the thoroughness of commissioning on completion, monitoring in use and subsequent maintenance. These factors fall into four groups:

- quality of the briefing and design process, including reliability of information used;
- quality of construction and commissioning process;
- quality of products and systems used; and
- quality of use of the building, and operation of its installed systems.

Achievement of faultless building is not easy. Features that differentiate building from manufacturing are listed in Box 4.1. It shows that the design–construction process is subject to many, possibly conflicting, constraints. Some result from incompatible requirements of owners and users; some from inconsistent provisions embodied in the regulations of different authorities. Some arise from the contractual relations of the various practitioners involved, and from imperfections in communication between the client and his agents, and between other participants in the process. Some arise from the fact that buildings have many uses at different times in their life, not all of which may have been or can be identified by the client in his brief, or can be foreseen by the architect during design work. Some are not apparent until users take up occupation. Furthermore, many buildings last for decades, if not centuries. Over the years they are subject to changes in ownership and to different, often unpredictable, cultural, economic and social influences.

BOX 4.1 CONSTRUCTION'S SPECIAL FEATURES

- Designing and building are usually separate activities, each practitioner – directly or through a third party – having a separate contractual responsibility to the client, with design work usually more or less complete before the builder is chosen. Even when both tasks are entrusted to one firm, work may be split between professions, with much work on site subcontracted.

- Most buildings are 'one-off' products, erected on ground that, even on a single site, may vary in character every few metres. Testing of prototypes is rare. Even when standard designs are used, details are frequently modified to satisfy site, regulatory or client requirements.

- Manufactured materials, components, assemblies and mechanical equipment may have been tested and quality assured in factory; but once brought on site they are likely to be handled, stored, assembled and installed under adverse weather and other conditions. Even when quality-assured components are used, and care is taken in their handling and installation, they may prove incompatible with their neighbours, the resulting chemical or mechanical interactions being a latent source of trouble.

- Construction workers move from site to site, changing employers from one job to the next. Types of work change as a scheme progresses as well as between jobs, as do size and skills required from the workforce. Employer relations change, as do coverage, expertise and quality of inspection and supervision. Quality of workmanship required from individual operatives is unlikely to be defined clearly.

- Buildings last for decades, more often than not for centuries, and parts of a building may have to be replaced at various times, receiving varying degrees of care, maintenance, repair and alteration during their life.

- Consequences of a defective design, selection of unsuitable component or material, careless installation, inappropriate maintenance or repair, and misuse during occupation may remain latent for many years, only showing up to cause trouble following an exceptional 'overload' such as a windstorm, earthquake or gas explosion.

- Technical requirements of regulations implicitly assume that a building will remain for ever as built; and the law tends to place all but timeless responsibilities for good performance on the original designer and producer in some countries.

- Statutory authorities regulate design and construction in many ways and stages during the process, and their requirements may be of a detailed, prescriptive or a functional and general character, and may be followed by examination and, possibly, formal approval of the resulting work.

- Supervision and inspection on construction sites are not usually systematic. Site testing of work in progress is rarely undertaken – except for certain civil engineering activities or when substandard work is discovered. When it is, rectification is likely to be costly and completion delayed.

- When defects are discovered, remedial work is unlikely to be easy for more than one reason: it may be difficult to determine the cause of failure, and a wrong diagnosis could well aggravate the problem; and neither the original work nor the changes resulting from the remedial work are likely to be properly recorded systematically.

- Environmental and user conditions vary within a single building, so the identification of defective components may prove troublesome; and, as the building is likely to be occupied, remedial work will be difficult to organize.

- Because responsibilities of participants in the process of design, manufacture, assembly and supervision are complex and sometimes ill-defined in contract, when latent defects are discovered it may be necessary for an owner to start litigation to recover the cost of resulting damage. Court procedures then take precedence over unbiased and open fact-finding. Consequently feedback to other designers and producers is restricted.

- While in traditional construction a degree of robustness and structural redundancy were the norms, this may not be so under new, possible cost-competitive conditions. A better understanding of how structures perform has enabled designers to work closer to 'limit states' for reasons of efficiency and economy – and, possible, as displays of technical skill. The traditional safeguards that protected practitioners of average competence are weakened.

- In offices where the partners and managers are experienced and their technical and professional staff possess above-average skills, and where there is effective quality management, risks may be taken. But, in the hands of the less experienced, serious troubles may result even when products of good quality are used. Lastly, it is not always easy for a client who builds infrequently to identify which firms are experienced, or possess above-average skills for a particular task.

There are national variations in the influences of such features on the construction process. In France, for example, the responsibilities of constructors to clients, their nature and time limits, are set out in the Civil Code, and contract procedures – at least for public works – are codified. Design checking, site responsibilities and handing over on completion of work are subject either to requirements of damage and responsibility insurance, as in France, or to a hierarchy of building officials, as in Federal Germany. Furthermore, there may be comprehensive arrangements for collecting and analysis of damage reports from experts called in to assess and adjudicate on insurance claims, as under France's Agence Qualité Construction's *Syncodés Information* system.

The UK is fortunate in the strength and experience of its national building research establishment, BRE, and in the standing of its professional institutions and its national standards body BSI, where the special needs of construction are recognized in its organizational structure. Furthermore, the practitioner has at his disposal a wealth of technical information and guidance from professional and technical organizations unmatched elsewhere.

That this wealth is not always tapped is suggested in the findings of the BRE investigation, described in Chapter 3, in the course of which the investigators could identify published information and guidance that, if used, would have eliminated all or almost all of the 400 or so kinds of design faults recorded. The French Agence pour la Prévention des Désordres also found that faults reported often arose from failures to take account of the state of art relating to everyday matters, such as the arrangement of steel reinforcement in a manner that obstructs easy concrete placement.

But faults also arise in situations where, to achieve an innovatory or an above-average architectural or economic solution to a design problem, risks are taken that call for above-average competence. Where it does not exist, serious troubles or inferior architecture may result, even when products of good quality are used.

Every year very substantial sums of money and much practitioners' time is spent putting right faults that should never have occurred. Some faults should have been discovered during the progress of a contract by a supervisor, clerk of works, site architect or engineer. Some are corrected during maintenance work. Some are not discovered until a strong wind, long period of drought, heavy rainstorm, act of arson or other natural or manmade event puts an additional stress on a building or a key component. Others are only discovered when a detailed structural survey is initiated by a change of ownership or management responsibilities. When the fault is discovered before associated work has been completed, correction may be easy. But later a fault is likely to be more difficult to correct, especially if it has detrimentally affected associated work, or the building is in use. At the worst, a fault may lead to a latent defect, claims for negligence, legal costs and much management time wasted preparing and giving evidence.

Obviously, recourse to an adequate, relevant collection of technical guides and technical specifications will not in itself achieve faultless construction; but it will go a long way. First place must be given to the documents issued by standards-making bodies, for two reasons:

- Reference to standards is becoming central to the regulatory process, as well as to the legal specification of client/customer technical requirements from producer/supplier.
- Replacement of national by European standards, which is a basic feature of 'New Approach' European legislation, should provide an opportunity to further the achievement of higher standards of quality in European buildings.

4.2 STANDARDS AND STANDARDS-MAKING IN THE UK

Standard: Document, established by consensus and approved by a recognized body, that provides, for common and repeated use, rules, guidelines, or characteristics for activities or their results, aimed at the achievement of the optimum degree of order in a given context.

BS 0: Part 1 1991, based on BS EN 45020: 1991

Standards are of key importance in achieving faultless building. Drawn up by all those with a particular interest in the subject – consumers, government agencies and departments, manufacturing and service industries, professional users and research organizations – they are used in legislation as guidance in meeting technical requirements; and reference to standards is one of the main ways of specifying technical requirements in contract and other documents, and of confirming through specified test procedures that specified requirements are satisfied. To achieve construction quality, therefore, an understanding of standards and standards-making is important.

Guidance on British Standards and standards-making is given in BS 0: *A standard for standards*, the form and content of which is shown in Box 4.2. Based on the European definition given in BS EN 45020: 1991: *Glossary of terms for standardization and related activities*, BS 0: 1991 defines a 'Standard' as a

document, established by consensus and approved by a recognized body, that provides, for common and repeated use, rules, guidelines or characteristics for activities of their results, aimed at the achievement of the optimum degree of order in a general context,

replacing the 1981 longer, but more user-friendly definition:

A technical specification or other document available to the public, drawn up with the cooperation and consensus or general approval of all interests affected by it, based on the consolidated results

BOX 4.2 BS 0: A STANDARD FOR STANDARDS

The standards-making process is set out in BS 0: *A standard for standards*, the most recent edition being published in 1991.

BS 0: Part 1: 1991 *Guide to general principles of standardization*
General aims and principles of standardization; the role and status of standards within the legal framework.

BS 0: Part 2: 1991 *Guide to BSI committee procedures*
Origin and objects of BSI, and organization and procedures governing the structure and constitution of committees, preparation and maintenance of standards, and UK involvement in European and international standards work.

BS 0: Part 3: 1991 *Guide to drafting and presentation of British Standards*
Presentation, arrangement and drafting of British Standards. Information on special considerations to be taken into account in drafting and defines details of style and typography.

Standards for building and civil engineering PD 6501: 1982

In parallel with revisions to BS 0, guidance on the preparation of British Standards for building and civil engineering was issued in 1982. Among other matters, PD 6501 looked at codes of practice, identifying two types, design codes and practice specifications, and dealt with the relationship between practice and product specifications. There was guidance on quality and grades, on basic data in standards, and, in Part 2, on presentation.

of science, technology and experience, aimed at the promotion of optimum community benefits and approved by a body recognized on the national, regional or international level.

Both place importance on approval by a body eligible to be the national member of the corresponding international and regional standards organizations – a description that suggests that standards-making must be a monopoly, and that at national, European and international levels, duplication by other bodies is deprecated.

Standards are of more than one kind:

- specifications of characteristics and performance of products, processes and systems;
- specifications of methods of evaluating performance, measurement, testing, sampling and analysis;
- glossaries, symbols, classifications; and

- codes of practice, guides, recommendations on product or process applications that bring together the results of practical experience and scientific investigation – now numbered in the General Series.

Specifications are detailed statements of a set of requirements to be satisfied. They should either include, or make reference to, methods of verifying compliance. Codes of practice bring together the results of practical experience and scientific investigation, and are thus recommendations on good accepted practices usually followed by competent practitioners. Availability and use in a design office of codes related to the office's sphere of work are demonstrations of that office's capabilities.

Standards have a number of uses:

- communication between interested parties of technical requirements and procedures needed to ensure compliance;
- protection of consumer interests through specification of adequate and consistent standards of quality for goods and services put on the market;
- promotion of safety, health and protection of the environment;
- promotion of economy in human effort, materials and energy in the production and exchange of goods; and
- promotion of trade by removal of barriers.

As standards become used increasingly in legislation and in contracts, BSI has found is necessary to explain clearly the status of British standards in law. A standard may be:

- cited in technical and trade descriptions of products;
- cited in commercial contracts;
- referred to in regulations – as a mandatory requirement, or as a deemed-to-satisfy example.

It has also been necessary to differentiate between the three ways in which reference to standards can be made:

- by general reference;
- by undated reference; and, as is the usual and preferred practice in the UK,
- by dated reference.

The national status of BSI was acknowledged in the publication of the Government White Paper *Standards, quality and international competitiveness*, in which, belatedly, the importance of standards, product certification and quality assurance for British trade overseas was recognized in Whitehall. To give support to standards work, the White Paper committed Government to a greater use of BSs as a basis for legislation and in public purchasing.

4.3 CONSENSUS AND THE GENERALLY ACKNOWLEDGED STATE OF THE ART

Standards are established by consensus, defined in BS 0 : Part I : 1991 as

> general agreement, characterized by the absence of sustained opposition to substantial issues by any important part of the concerned interests and by a process that involves seeking to take into account the views of all parties concerned and to reconcile any conflicting arguments,

but not necessarily unanimity.

An issue in standards-making is how to ensure that standards, as consensus documents reflecting the generally acknowledged state of the art in their field, do not inhibit innovation or, as bad, allow practices to continue and products to be put on the market that advances in a medical or other science, a more through analysis of safety or health records, regulators' and users' experience, or advances in manufacturing technology demonstrate are unacceptable.

When work is started on a new, or revised standard, there is an announcement in the monthly *BSI Update*. The initial draft may be entrusted to a small panel or an outside consultant. For certain drafts there are arrangements for Government to support the work under a BSI consultancy scheme administered, in the case of work for construction, through a Department of the Environment department or agency.

After consideration, and possible amendment in committee, the draft is issued 'for public comment', again noted in *BSI News*. If there is consensus among comments received and in the committee, a final draft is prepared and processed for publication. If comments received show lack of consensus, the committee may review the scope or even the need for the standard. There may be differences of view within the committee; which, if the chairman is unable to resolve, may have to go on appeal to the main standards committee or, for construction, B/– Standards Board for Building and Civil Engineering. This does not generally happen but there are rare occasions when a committee, having a particular interest or opinion, can seriously delay drafting, or alternatively lead to a compromise document.

To achieve consensus in standards-making:

- A standard should be wanted by all the parties involved.
- It should be used, which means that the intended application should be clearly understood when work is started on its preparation.
- Application depends in the first place on the voluntary action of the interested parties.
- The standard should be concise, clear and unambiguous, taking into account that a standard may, by reference, be given legal recognition in contracts or through legislation.

- As a standard may contain options, these need to be clarified before use in contracts or legislation.

Where it has not yet been possible to issue a standard, its precursor may be published as a Draft for Development. BSI used to number Codes of Practice in a separate CP series. Now, as they are revised, they are numbered and listed in the General Series. Only PDs (Published Documents) – mainly guides and reviews of design methods etc. – and DDs (Drafts for Development), with a small number of specialist specifications, are listed in separate series.

4.4 USER INTERESTS

User interests are promoted through adoption of procedures that ensure adequate and consistent quality of goods and services, and at the same time benefit the whole industry rather than a single producer or group of producers. Following a 1989 review of consumer participation, a new Consumer Policy Committee OC/11 was set up in June 1992 with representation from many different consumer interests. For the construction user, a Construction User Group was set up during 1993. The Group, whose membership reflects a wide cross-section of user interests, operates under the aegis of OC/11.

BSI is doing much to help users of British Standards to overcome such problems as:

- whether there is a standard relevant to a matter in hand, and whether other standards are also relevant. Here the much improved BSI Catalogue Subject Index is useful, but as yet not infallible. There is the BSI Library Enquiries Service, now at Chiswick in West London, to fall back on. Publication by BSI from time to time of Handbooks, like Handbook 3: *Standards for building* and Handbook 22: *Quality assurance* is another way of finding relevant British Standards.
- whether there is an equivalent European or international standard. Here guidance is given in the BSI Standards Catalogue. The sign ≠ denotes that there exists a related but not equivalent international or European standard, the sign ≡ denotes an equivalent standard in every respect, and = denotes a standard that is technically equivalent but worded and presented differently.
- how to keep up to date on standards, a problem that can be tackled by users in one of three ways:
    - by joining PLUS, the BSI Private List Updating Service tailored to an individual's or organization's needs;
    - by subscribing to the monthly BSI Catalogue updating service; and
    - by scanning monthly *BSI Update*. A number of professional and trade journals report new and amended British Standards, but reporting may not be comprehensive.

At an international level, BSI is an active participant in the work of the ISO Consumer Policy Committee, a working group that will be publishing in 1995 an updated ISO/IEC Guide 37: *Instructions for use of consumer products*. This explains how to produce good instructions, being directed at standards-makers, product designers, manufacturers and technical writers. It also explains how instructions for use should be translated from one language to another; the design of warning notices; the role of colour; and it includes, like BS 4940, the need to include advice on disposal of wastes.

The voluntary nature of BSI work can, on occasion, also cause problems for users:

- Work started will have been reported in *BSI Update*, and may be mentioned in the technical press, but a firm or group affected by the work may not have kept abreast with notices, or obtained drafts for comment when issued. The group's internal arrangements may be inadequate to ensure that comments are made in time.
- Although the organization is represented on the committee, its representative may not keep it sufficiently well informed of progress, or have sought an adequate brief, leaving the organization unaware of the effects on its members until too late to influence the drafting.

Cost of attendance at BSI meetings in the UK is borne by individuals, their firm or professional or industry association. However, there are now special arrangements for meeting the expenses of individuals representing BSI at European standards meetings. Because the interests of a professional or consumer group may not be, or be inadequately represented in BSI work, members serving in a personal capacity on the Consumer Policy Committee and its four Coordination Committees may draw on the DTI Consumer Travel Expenses Fund.

To clarify the respective roles of BSI and the British Board of Agrément, since 1983, by mutual agreement, BBA has the task of assessing 'materials, products, systems and techniques' for certification when there is no relevant BS, or where there is an innovative element resulting in as good or better performance. It will not renew or maintain certification when there is an appropriate relevant BS. BBA and BSI jointly agreed:

- to use BS 5750 (now the ISO 9000 series) as a common basis for assessment and surveillance; and
- to use, wherever possible, NATLAS/NAMAS-accredited testing laboratories.

4.5 BRITISH STANDARDS AS AIDS TO QUALITY IN CONSTRUCTION

In one way or another almost all British Standards for design and construction of works, and the properties and performance of products and systems used, are directed at quality in construction. Three kinds of British Standards, directed at processes rather than products, are highlighted in this and the following chapter. The first group, to which reference has already been made, is about information transfer, and particularly transfer of information about construction products and services; the second group, listed in Box 4.3, is directed at workmanship on site. The third group, about quality systems and the processes of managing design, manufacture and construction, is the subject of Chapter 5, 'Achievement of quality through management'.

Workmanship on site is but one of many factors affecting construction quality, although it is the one most likely to be seen, and commented on, by clients and the lay public. Guidance is available in many publications, including British Standard Codes of Practice. In 1989–1990, BSI brought together this guidance in BS 8000: *Workmanship on building sites*. Issued in 15 Parts, the Standard has been criticized for its cost rather than for its comprehensiveness. Box 4.3 lists the topics covered.

4.6 EFFECTIVE INFORMATION TRANSFER: STANDARDS AS AIDS TO COMMUNICATION

There is general agreement that improved information flow is necessary to overcome the fragmented organization of the building process.

> T.J. Griffiths, UK Building Research Station

The goal of communication is exchange of information. The general interest in communication problems in the building industry of advanced countries indicates that the present system of communication is not satisfactory.

> H.P. Sundh, Norwegian Building Research Institute
> CIB Congress 1971: 'Research into practice:
> the challenge of application'

A key factor in achievement of construction quality is the efficient and effective communication of technical information between clients and designers, designers and builders, and installers of equipment and its users. Communication of reliable, appropriate and easy-to-handle information from manufacturers of construction products and suppliers of services to specifiers and purchasers is of equal importance.

C.R. Honey, in BRS Current paper CP4/89, *Architectural design*, reviewed, as part of a wider study of coding and data coordination in

BOX 4.3 BS 8000: 1989–1990 *WORKMANSHIP ON BUILDING SITES*

Part 1   Code of practice for excavation and filling
            Recommendations on basic workmanship
Part 2   Code of practice for concrete work
            Section 2.1 Mixing and transporting concrete
            Section 2.2 Sitework with *in situ* and precast concrete
Part 3   Code of practice for masonry
            (Covers tasks frequently carried out in relation to brick
            and blockwork. Does not cover stonework.)
Part 4   Code of practice for waterproofing 1989
            (Covers tasks carried out in relation to waterproofing in
            tanking, damp-proofing and roofing applications.)
Part 5   Code of practice for carpentry, joinery and general fixings
Part 6   Code of practice for slating and tiling of roofs and
            claddings
            (Applies to the laying and fixing of clay and concrete
            tiles, nature and fibre-reinforced slates and their associ-
            ated fittings and accessories.)
Part 7   Code of practice for glazing
            (Does not cover off-site glazing; includes specialist
            glazing techniques, roof glazing, glazing of furniture and
            fittings and use of profiled glass and glass blocks)
Part 8   Code of practice for plasterboard partitions and dry
            linings
Part 9   Code of practice for cement/sand floor screeds and con-
            crete floor toppings
Part 10   Code of practice for plastering and rendering 1989
Part 11   Code of practice for wall and floor tiling:
            Section 11.1: Ceramic tiles, terrazzo tiles and mosaics
            1989
            (Applies to the fixing of ceramic tiles and mosaics to
            walls, floors and to the fixing of terrazzo tiles to floors)
            Section 11.2: Natural stone tiles
            (Covers granite, marble, travertine, slate, quartzite, lime-
            stone and sandstone)
Part 12   Code of practice for decorative wall coverings and paint-
            ing
Part 13   Code of practice for above-ground drainage and sani-
            tary appliances
Part 14   Code of practice for below-ground drainage
Part 15   Code of practice for hot and cold water services (domes-
            tic scale)

the construction industry, the handling of information by designers. He noted that

> Information techniques alone will not improve the quality of architectural design. Very much depends on the architect's skill in the handling of information in the service of his conceptual skills.

Nevertheless, he is helped by these techniques, particularly:

- presentation of requirements in an ordered way; and
- presentation of information on products in an ordered manner with relevant properties for each stage of design readily available.

Much the same conclusion was reached by a team from the University of York's Institute of Architectural Studies in a report on *Design decision making in architectural practice: the roles of information, experience and other influences in the design process*. One source of information studied was trade literature. Box 4.4 lists British Standards that serve as aids to better communication.

BOX 4.4 BRITISH STANDARD AIDS TO BETTER COMMUNICATION

Selected list of British Standards aimed at improvement of communication between participants in the design–construction process

Glossaries
- BS 4422: 1975–1990 *Glossary of terms associated with fire*
- BS 5408: 1976 *Glossary of documentation terms*
- BS 6100: 1989–1991 *Glossary of building and civil engineering terms*

Guides and manuals
- BS 3700: 1988 *Recommendations for preparing indexes to books, periodicals and other documents*
- BS 4884: 1973–1983 *Specification for technical manuals*
- BS 4940: 1993 *Technical information on construction products and services*

  Part 1: Guide to initiation and commissioning; 2: Guide to content and arrangement; and 3: Guide to presentation (Headings for the arrangement and presentation of technical information are based on the 1993 CIB Master List)
- BSI PD 6501: Part 2: 1984 *The preparation of British Standards for building and civil engineering*

  Part 2: Guide to preparation

Table 4.1 CIB Master List of headings for arrangement and presentation of information in technical documents for design and construction

Heading	Information given under heading
0 Document	Title of document; originator; publication details
1 Identification Brief description	Range of products or services covered; proprietary/trade name; manufacturer/supplier; identification information, e.g. material, intended use, finish, method of manufacture
2 Requirements	Requirements that the product of service will meet, such as technical specifications, regulations and standards
3 Technical description	Intrinsic properties, e.g. composition, size, mass, colour
4 Performance	Behaviour of product or service in use: structural; fire; resistance to water, chemicals, mould etc.; thermal, optical, acoustic, electrical; resistance to attack; service life, durability, reliability
5 Design work	Technical and economic suitability; design methods and calculations; limitations and precautions; model specification clauses; examples of design details
6 Sitework	Handling, storage, installation, fixing, cleaning, protection and other information of direct interest to builder
7 Operation	Information for building user, including operation of components such as blinds, windows and security devices; commissioning and operation of services and equipment
8 Maintenance, repair, replacement, disposal	Information required, after installation or completion of work, on cleaning, maintenance, servicing, repair, replacement and disposal of used product
9 Supply	Packaging, transport and delivery; prices, conditions of sale and other commercial and contractual information
10 Manufacturer/supplier/ importer	Information about manufacturer/supplier/importer's administrative and technical organization
11 References	Related publications, e.g. test reports and installation instructions; reference to other publications with addresses of manufacturers/ suppliers of associated products and services; locations where examples of installed work can be inspected

BS 4940: 1993 *Technical information on construction products and services* is in three parts. Part 1 is intended for those in industry who have overall responsibility for an organization's publication policy. It deals with the preparation of the brief for technical publications, and covers commissioning of technical writers and designers of technical publications. Part 3 is intended for those concerned with the presentation of technical information, covering such matters as typography, graphics and page layouts.

Part 2 is of wider interest. It gives guidance on the content and arrangement of technical information, the aim being to ensure that technical information is complete, of good quality, reliable, well structured, easy to retrieve and, where possible, authenticated by reference to standards or equivalent documents. It is planned to transform the standard into a form usable internationally as an ISO document, or similar documents.

The headings for technical information in Part 2 are taken from the 1993 edition of CIB Report 18, *CIB Master List of Headings for the Arrangement and Presentation of Information in Technical Documents for Design and Construction*, published by the General Secretariat of CIB, the International Council for Building Research, Studies and Documentation, Post Box 1837, NL-3000 BV Rotterdam.

The CIB Master List (Table 4.1) was first published in 1964 as Report 3: *Master List of Properties for Building Materials and Products*. New editions were published in 1970, and in 1972 when *Lists for Building Elements and Services* were added. In 1993 wider uses were recognized and, as CIB Report 18, the Master List was extended to cover technical documents for design and construction in general. At the same time the headings were linked with terms used in 89/106/EEC essential requirements, and associated interpretative documents. The CIB Master List is, therefore, an internationally agreed list of headings for the arrangement and presentation of information used in design, construction, operation, maintenance and repair of buildings and building services, and in associated documents on the supply of construction products and services, their manufacturers and suppliers. Its arrangement follows the process of design, construction, operation, maintenance, repair, and supply, enabling technical documents to be arranged in a consistent form.

FURTHER READING

The importance of standards in the achievement of construction quality is a central theme of this book. In the UK, the British Standards Institution is the national standards-making body. BSI also operates an accredited testing laboratory, and quality assurance and product certification services. The history and organization of BSI as a standards-making body are described in Chapter 11, section 11.6.2. In 1994 BSI Head Office, BSI Standards and most BSI Sales and

Customer Services, including BSI Information and Library Services, were relocated in new, specially equipped facilities at Chiswick in West London. Details of the new BSI organization – locations, addresses and phone and fax numbers – are given in Chapters 15 and 17.

BS 0: *A standard for standards* is a guide to the general principles of standardization; to BSI committee procedures; and to the drafting and presentation of British Standards. The monthly *BSI News*, *BSI Update* and the annual *BSI Standards Catalogue* update information on BSI's activities. The *Catalogue* lists public and other libraries in the UK where sets of British Standards are kept. It also lists all British Standards, most with short notes on their scope and content.

The importance of effective transfer of information between participants in the design–construction process is stressed in many publications. It has been the concern of two CIB Working Commissions: W57 Building documentation and information transfer, and W74 Information coordination in the building process. C.R. Honey in *Architectural design* (BRS Current paper 4/69), a contribution to an early study of coding and data coordination for the construction industry, reviewed how information is handled in the design process: a matter taken up later by M. Mackinder and H. Marvin in *Design decision making in architectural practice: The roles of information, experience and other influences during the design process*, Research Paper 19, Institute of Advanced Architectural Studies (University of York, 1982). Other publications include: J.J.N. O'Reilly, *Better briefing means better buildings*, Report BR 95 (BRE, 1987); D.T. Crawshaw and K. Snook, *Production drawings – arrangement and content*, Information Paper IP 3/88 (BRE, 1988); and *Project information for statutory authorities*, Digest 271 (BRE, 1983).

Two more recent publications are: CIB Report 18, *CIB Master List of Headings for the Arrangement and Presentation of Information in Technical Documents for Design and Construction* (CIB General Secretariat, Post Box 1837, NL-3000 BV Rotterdam, 1993); and BS 4940: 1994 *Technical information on construction products and services:* Part 1 *Guide for management*, Part 2 *Guide to content and arrangement*, Part 3 *Guide to presentation*.

On aids to better communication, advice is given by Ron Brewer in a BRE Occasional Paper, *Write it right: a guide to preparing technical literature*, published in September 1994. It lists a number of other publications, mainly on technical writing.

Achievement of quality through management

5

Quality is the sum of:

- knowing customer's needs
- designing to meet them
- faultless construction
- reliable bought-in components and assemblies
- certified performance and safety
- suitable packaging
- punctual delivery
- effective back-up services
- feedback of field experience

National Quality Campaign 1983

5.1 QUALITY SYSTEM STANDARDS AND QUALITY ASSURANCE

The contribution, in general, of standards to the achievement of faultless building was examined in Chapter 4. Their application in different management processes, including design and construction management, is discussed in this chapter. As background to the application of quality systems, there is first a description of how quality standards and quality assurance came about in the UK.

In 1962 a National Council for Quality and Reliability (NCQR) was set up under the wing of the British Productivity Council. Supported, until 1971, by a government grant, NCQR represented the UK at meetings of the European Organization for Quality Control. In 1967 the Council organized, with the Productivity Council, a Quality and Reliability Year. However, quality assurance received limited support from industry and the engineering professions, who did not regard quality assurance as an academic discipline, with its practitioners coming from a range of backgrounds in engineering and science.

Looking back on this period, the lack of initiative by Government and industry is surprising, particularly as third-party voluntary or mandatory certification was already well established in countries like Federal Germany, and was already causing barriers to trade for British exporters, who were losing or had lost the benefits of protected markets in the former British colonial territories. A study then in progress at the Building Research Station was also showing the increasing role of third-party certification of building products in building regulation in the Netherlands and Scandinavia, as well as in Germany.

In 1970 a committee chaired by Sir Eric Mensforth had expressed concern about national arrangements for the attainment of quality in the engineering industries. Set up two years earlier by the then Ministry of Technology, the Mensforth Committee sought the views of some 800 government, industry and research organizations. Having reviewed the inadequate quality assurance arrangements for statutory purposes and in relation to the promotion of UK exports, it recommended that a national agency – a Quality Board – be set up. The Board would be responsible for authorizing national certification bodies in defined fields.

The concept of a board was to prove to be in advance of general thinking in Government and industry, and had a mixed reception. In particular, the Confederation of British Industry did not favour the concept, fearing that the Board would have to have 'extensive, legislative and judicial authority'. However, there was one fruitful result of this early initiative: a strengthening of interest within the British Standards Institution. In 1971 its Mark Committee, until then mainly concerned with the BSI Kitemark scheme, became the Quality Assurance Council and, through a supporting structure of BSI committees and panels, did much to promote an interest in quality assurance for standards work, an interest that was to culminate in the 1977 report by Sir Frederick Warner, *Standards and specifications in the engineering industries*. It also saw the issue of BS 4778: 1971 and, a few years later, BS 5179: 1974 *Guide to the operation and evaluation of QA systems*.

The 1978 consultative document was to take up again the Mensforth concept of a British Quality Board, suggesting that

> a completely new body would have the advantage that its composition and terms of reference could be precisely tailored to the task envisaged and that at the same time demonstrate a clear national commitment to the importance of quality.

It was suggested that the scale of effort 'in terms of secretarial, promotional and supporting activities might be built up to about £500 000 per annum'. The publication by the former Department of Prices and Consumer Protection in 1978 of a consultative document, *A National Strategy for Quality*, can be said to have marked a turning point in quality assurance in the UK.

Support was given to this proposal by the Advisory Council for Applied Research and Development (ACARD) in its 1982 report *Facing International Competition*, which recommended as follows:

- Government should establish a national accreditation scheme for certification and approval bodies.
- Government, with the Confederation of British Industry, trade associations, and the National Economic Development Council through all its sector working parties, should encourage the use of and publicize the benefits of quality management systems based on BS 5750.
- Public and commercial purchasers should make use of independent schemes for assessing the quality management procedures of their suppliers in preference to carrying out their own assessment.
- Individual firms and trade associations should take steps to establish many more product certification and approval schemes, perhaps on a sectoral basis, particularly for internationally tradeable products where quality is important.
- Government should adopt a preventive approach to the enforcement of safety requirements in the UK, making use of independent certification and approval schemes wherever possible.

5.2 BS 5750 AND THE BS EN ISO 9000 QUALITY SYSTEM SERIES OF STANDARDS

Not all the ACARD recommendations have been accepted fully by the UK Government. Although BSI published the first national standard for quality systems, BS 5750, in 1979, and the standard has been adopted as a European standard, EN 29000 and, internationally, as ISO 9000, government departments in their privatization programmes have not insisted that the new organizations and firms make use of quality system standards for the development of which the UK was the pioneer. It is to be hoped that the European Commission's recommendation that the ISO/EN series of quality system standards be adopted as an integral part of signifying conformity to technical specifications will strengthen departments' attitudes to the application of quality systems in the public sector and associated manufacturing and service industries.

Drawing on five years' experience, during 1993–94 the BS 5750 series of quality systems standards has been reviewed and, where necessary, updated to incorporate best management practice and give clearer management responsibilities. Reflecting their European and international status, the British series has been given a new set of numbers, and titles changed from 'specification' to 'model'.

There have been no changes in the basic structure, the aim of any changes being to improve presentation of requirements, making the set of standards more user-friendly, and to ensure consistency among the set and keep the series stable to facilitate continuity, changes include:

- As well as giving attention to customers' needs, quality policies should be directed at organizational goals, and take notice of customers' expectations.
- Responsibilities are extended to active prevention of product problems.
- There is a requirement that management reviews be carried out by executive management, and should take account of stated policies and objectives.
- Controls described in documented procedures should include purchased services and selection of subcontractors, whose nomination should be evaluated.
- Systems should include arrangements for verification of conformity to requirements at a supplier's premises.
- BS EN ISO 9002: 1994 now includes servicing.
- Under a related ISO 1001 series, there are guidelines for the auditing of quality systems.

The BSI's *QA Guide to the 1994 Revision of BS 5750* explains the changes, which have the following aims:

- To clarify the principal quality concepts and provide guidance on the selection and use of the series for both internal quality management and external assurance purposes.
- To provide guidance on quality management and quality system elements, describing a basic set of elements by which quality systems can be developed and implemented for internal quality management purposes.
- To provide a model for design/development, production, installation and servicing quality systems for use where a contract between two parties requires that a supplier's capability to design and supply a product or service can be demonstrated in a situation where the customer/client/purchaser requirements are set out in terms of performance required, the supplier being responsible for interpreting requirements in a design and specification, and undertaking any necessary development work before manufacture, and for manufacture, whether undertaking in house or through subcontracts, including installation and field trials.
- To provide a model for production and installation for use where there is a technical specification for the product or services, usually provided by customer/client, and where a demonstration of the supplier's capability to control the processes that determine acceptability of the product or service supplied is required.
- To provide a model where the requirement is limited to quality systems for final inspections and tests demonstrating a supplier's capability to detect and control any product nonconformity during final inspection and test.

BOX 5.1 THE 1994 BS EN ISO 9000 SERIES: *QUALITY MANAGEMENT, QUALITY ASSURANCE AND QUALITY SYSTEMS STANDARDS* (FORMERLY BS 5750)

- BS EN ISO 9000-1: 1994 *Guidelines for selection and use*
  Clarifies the principal quality concepts and provides guidance on the selection and use of the ISO 9000 family of standards on quality systems, which can be used for internal quality management purposes and for external quality assurance purposes.

- BS EN ISO 9001: 1994 *Quality systems – Model for quality assurance in design, development, production, installation and servicing*
  Specifies quality requirements for use where a supplier's capability to design and supply conforming product needs to be demonstrated.

- BS EN ISO 9002: 1994 *Quality systems – Model for quality assurance in production, installation and servicing*
  Specifies quality requirements for use where a supplier's capability to supply conforming product to an established design needs to be demonstrated.

- BS EN ISO 9003: 1994 *Quality systems – Model for quality assurance in final inspection and test*
  Specifies quality requirements for use where a supplier's capability to detect and control the disposition of any product non-conformity during final inspection and test needs to be demonstrated.

- BS EN ISO 9004-1: 1994 *Guidelines*
  Describes a basic set of elements by which quality systems can be developed and implemented for internal quality management purposes. Selection of appropriate elements and the extent to which these elements are adopted and applied by a company will depend on such factors such as market being served, nature of product, product process and consumer needs.

The ISO 9000 series is to be kept up to date through a two-phase revision. The first phase took place in 1994; the second phase will take place, at the earliest, after five years, i.e. not before 1999. The 1994 revised series is set out in Box 5.1. How the up-dated series relates to the earlier BS 5750 series is shown in Table 5.1.

Table 5.1

Formerly known as	Now known as
BS 5750 Part 1: 1987	BS EN ISO 9001: 1994
BS 5750 Part 2: 1987	BS EN ISO 9002: 1994
BS 5750 Part 3: 1987	BS EN ISO 9003: 1994
BS 5750 Part 0.1: 1987	BS EN ISO 9000–1: 1994
BS 5750 Part 0.2: 1987	BS EN ISO 9004–1: 1994

To complement ISO 9000, the ISO standards listed in Box 5.2 are directed at the auditing of quality systems.

BOX 5.2 AUDITING QUALITY SYSTEMS

ISO 10011   *Guidelines for auditing quality systems*

ISO 10011-1   *Auditing*

ISO 10011-2   *Qualification criteria for quality systems auditors*

ISO 10011-3   *Management of audit programmes*

ISO 10012-1   *Management of measuring equipment*

ISO 10013   *Developing quality manuals*

The NAMAS Executive, which has been using Guide 25 as a means by which testing laboratories can demonstrate technical competence and that they operate a quality management system, has, in its spring 1994 *Newsletter* (Issue No 8), compared in detail some of the differences between these two procedures, stressing that users of test data should be aware of these differences. Basically, EN 45001: *General requirements for the operation of testing laboratories* uses the principle of 'peer review'. It not only requires assessment of compliance with a laboratory's quality system elements but also its technical capability. Assessment by a certification body under ISO 9000 only addresses itself to an organization's capability in terms of its quality management systems. It does not extend to the organization's specific technical capability.

5.3 QUALITY ASSURANCE IN DESIGN AND CONSTRUCTION

Application to quality management systems in building design and construction, and design and installation of building services, has required some compromises but is now well established. BS 5750's first applications were largely in manufacturing industry although, more recently, the series has been widely used in service industries, in the public sector and in the construction industry and professions.

As studies like the BRE study of quality in traditional housing have shown, up to half the faults observed could be attributed to the design and sitework stages of a project, or to communication between designer and constructor. Faults arising from the quality of materials if used properly are fewer, except where the specifier is misled by inadequate or incorrect technical information, or where an otherwise good product has been badly stored on site, misused or used in inappropriate situations.

If buildings are to be trouble-free, therefore, more attention needs to be given to applying quality assurance principles to design and sitework, including product selection and specification, and to supervision of their handling and protection on site. Now that there is up-to-date guidance, more attention needs to be given to an organization's competence in presenting, to specifiers and purchasers, accurate and complete technical information on the products and services that it offers.

The quality philosophy on which BS 5750 is based can be summarized in three maxims:

- Management should be quality conscious.
- Formal quality procedures should be used.
- All transactions should be documented.

While these maxims have universal application, their direct application to construction presents difficulties, especially in small firms and where much of the work is parcelled out into packages, for example, to specialist groups in design work and to subcontractors and specialist installers of building services.

There are, and always will be, misconceptions about quality assurance. Quality assurance is a management instrument whose central aim is to ensure that communication of requirements and other information is done in a structured and systematic manner, whether in house, between client and consultant, designer and constructor, contractor and supplier, or subcontractor. A central aim is to ensure that damage arising from inadequate, misunderstood or missing information is minimized; responsibilities are transparently clear; and decisions made are systematically recorded to facilitate corrective action rather than allocate blame. However, quality assurance only provides support when applied with understanding. Then it can make the complex process of design and construction more effective, and completed works – and not just the manufactured products – up to 'specification'.

Quality assurance is a management process to provide high probability that the objective of the product or service will be achieved. It is a systematic way of ensuring that organized activities happen in the way they are planned. It is a management discipline concerned with anticipating problems and with creating the attitudes and controls that prevent problems arising. It is common sense written down or, in short, the logical extension of current good management practice. But it is not a selection process to provide a suitable level of performance in the product or service on offer – that decision still lies with the designer or specifier.

Some of the issues not yet fully resolved include:

- the need to agree recommendations for conventions at interface points where there is communication with other parties so that information is given in a uniform way;
- standard presentation of information on enquiries and responses, invoice and delivery information from suppliers;
- agreement on product identification: marking on item and packaging, delivery documents etc., especially where there are different classes/grades of similar products, e.g. brick and concrete blocks;
- standard procedures for handling, storage, packaging and delivery including protection after final inspection and test;
- the need for stockists to keep records of product origins;
- procedures for designers to supply drawings;
- standard arrangements for pre-contract and site meetings;
- a clear convention on responsibility for examination and degree of checking of drawings, schedules and specifications;
- appropriate rules for display of information on QA logos, to avoid deliberate or accidental misrepresentation; and
- agreement on complaints procedures: failure to conform to specification; registration of acceptable deviations.

5.3.1 Interested parties in quality assurance for construction

Many parties have an interest in the achievement of quality and in the appropriate application of quality assurance:

- owners and users of buildings and civil engineering works, together with householders and public and private tenants;
- public authorities, central and local, with statutory responsibilities for public health and safety, and to a lesser degree amenity and convenience; also for the conservation of national resources like energy, fuel and water;
- groups with a financial interest in construction works: public agencies, commercial firms and private individuals; banks, building societies and investment institutions; firms and syndicates of insurers giving cover against fire and other damage risks and indemnifying designers, builders and others against claims for negligence under contract or tort; and
- the many participants in the construction process – designers, builders, installers of services etc., as well as the producers and suppliers of materials and equipment used – all having a double

interest in the achievement of quality: to satisfy clients and customers, thus promoting business and profit; and to limit avoidable costs, for example by reducing rejected and defective work.

Each quality assurance scheme is individual to the firm using it, and covers as much or as little of its operations as the firm wants. Having worked out a scheme, the firm may seek registration through a certification body, which will examine the scheme against any agreed sectoral application document. When the quality system has been assessed and accepted, the firm or organization, in using the registration body's logo and any description of the scheme, should refer to scope of the firm's operations covered by the scheme.

Box 5.3 lists matters to be considered when quality systems are introduced into design and construction.

Practical guidance on the auditing of quality management systems, based on the general experiences of certification bodies, is given in the ISO series 1011. Part 1: *Auditing* covers first-, second- or third-party audits, with clear specifications of the responsibilities of client, auditor and auditee. Part 2: *Qualification criteria for auditors*, as its title indicates, sets out minimum criteria and how to judge compliance by individual potential auditors with these criteria. Part 3: *Managing an audit programme* describes the activities that should be addressed when establishing audits of quality systems. Together the audit procedures set out in the three Parts should ensure that:

- documented quality system procedures are practical, understood and followed, and define the needs of the business to which they relate;
- training of personnel operating the system is satisfactory.

In the issue of two further groups of British Standards the scope of the quality system concept has been widened. The first covers what has come to be known as total quality management, a concept that recognizes that, for the efficiency of an organization, every part of its business should work towards the same goal. BS 7850: *Total quality management* looks at procedure, an organization's culture, management and work attitudes. It is published in two parts: BS 7850: Part 1: 1992 *Guide to management principles* and BS 7850: Part 2: 1992 *Guide to quality improvement methods*.

5.3.2 Economics of quality assurance

Quality assurance involves costs; it also involves benefits. The costs fall on a particular business; the benefits are shared between producers and customers or clients. Different certification bodies have different methods of charging; some have a published tariff, others state that their assessment service is available 'at an appropriate fee'. Fees include a proportion of the costs of NACCB accreditation of the certification body and its maintenance, possibly offset by initial grants

BOX 5.3 A GUIDE TO QUALITY SYSTEMS FOR CONSTRUCTION

- Coordination and monitoring must be in the hands of one man.

- The system must take into account all functions: design, manufacturing, subcontracting, storage, erection, installation commissioning, and particularly unusual client requirements.

- Site work instructions should be put on paper in simple form for operatives.

- Records are the objective evidence of meeting client's requirements, and need an efficient storage and retrieval system.

- When faults are discovered or defects are reported, they must be corrected by prompt and effective corrective action; action that must extend, where appropriate, to design faults, and faulty products and services provided by subcontractors. Corrective action should be recorded.

- Purchased material coming on a construction site is more likely to be checked against delivery documents for cost control than control of its quality. CE marking of products alone is likely to be of limited value as a mark of quality. Material supplied by the client must also be subject to control over quality.

- With latent defects liabilities becoming increasing onerous, handing over of completed works needs to be formalized, possibly modelled on the French *réception*, before the issue, for example, of a certificate of practical completion.

- The system should include control over disposal of 'condemned', i.e. non-conforming materials. Written control procedures are necessary to make it possible to establish quickly at all times whether material has been inspected, and approved or rejected.

- Procedures are required to protect and preserve product quality during handling and storage of materials on site, as BRE surveys have shown how wasteful and damaging handling and storage of materials can be on building sites.

- Periodic checks and systematic reviews are essential to maintain any quality system.

from the Department of Trade and Industry. There will be other costs, difficult to assess, such as initial consultancy fees and the costs of maintaining a quality system. There will be a cost in training quality management and control staff; in working with the certification body in preparing the necessary quality assurance procedures and assessment schedules; in preparing work for tests and their cost; and the cost of any consequent changes to operational practices needed. Further costs will possibly arise from keeping better records, marking products, providing certificates to customers etc. There will also be fees for periodical surveillance visits from the certification body.

A claim that can usually be substantiated is that the introduction of a quality system in manufacturing reduces the amount of material that has to be re-worked or disposed of as waste. In design and construction there should be less abortive or defective work. But not all such benefits are easy to substantiate. The introduction of a quality system in manufacturing should improve efficiency and possibly save manpower by encouraging better work practices; but it is not always possible to re-employ the production workers made redundant.

One area where benefits may be substantial is in the marketing of a firm's products or services, though it is not always easy to quantify the benefit in money terms. It may establish or reinforce the firm's reputation, and the quality and reliability of its products or services against competitors: a benefit less easy to substantiate, however, if all firms in a sector of industry introduce quality assurance. A further benefit may be protection against competition from importers of substandard goods. There may be parallel benefits for firms exporting goods or services. For a sector of industry or a professional group, the existence of quality assurance could enhance the sector's or group's standing, particularly where participation in a scheme is a requirement of membership. It may have special advantages where a competing group offers an alternative technique, product or service, or where competition from overseas has become serious.

Although such benefits are not easy to quantify, they should show up reasonably soon through the market, particularly where conditions are competitive, though conditions may exist where commercial buyers in an economic recession, and government departments and other public authorities subject to budgetary pressures, abandon quality for cheapness.

For the design professions and building and civil engineering contractors, the benefits of quality assurance may be even less easy to quantify until it results in a reduction in insurance costs, and a simplification of building regulation procedures. However, as larger public and commercial clients begin to require certification of competence as a condition of contract, firms wishing to tender will have to look on the costs of quality assurance as an essential overhead.

Benefits at a national level are more complex. To support the

National Quality Campaign, the Department of Trade and Industry has been making grants for specific quality assurance activities to promote greater industrial efficiency and support to UK firms competing in world markets. There should be a further benefit in that a widespread application of quality assurance could reduce the risk of dangerous and hazardous situations, allowing a degree of deregulation without an erosion in the standards of public safety.

BS 6143: *Guide to the economics of quality* attempts to give guidance on the costs and benefits of quality. Part 1: 1992 *Process cost model* is intended to be applied to a wide range of business; Part 2: 1990 *Prevention, appraisal and failure model* gives guidance on determining costs associated with defect prevention and appraisal activities, losses due to internal and external failures, and the operation of quality-related cost systems for effective business management.

Being fairly sophisticated, the guidance does little for smaller businesses, who have in recent years complained about the cost of some certification schemes. That certification bodies are starting to consider launching a 'Small Business Option', which would offer a similar status at lower cost, is therefore a move that NACCB as the watchdog should welcome.

5.4 QUALITY SYSTEM STANDARDS IN EUROPE

Compared with the UK, continental Europe was slow to introduce quality systems standards, at least in following the procedures set out in the international and European versions of BS 5750: however, guidance given by the European Commission at the end of 1992 on the use of European quality system standards (now consolidated into the EN ISO 9000 series) as part of the Union's conformity assessment policy gives weight to their wider adoption. The Commission points out that firms need to set up quality systems, thus enabling them to guarantee that the required quality will be obtained at the lowest cost; while customers, including public authorities, need to know whether their suppliers' quality systems provide the necessary guarantee of quality. Commission document DOC. CERT 92/9 spells out the Commission's policy. The relationship between the international and European quality systems standards is illustrated diagrammatically in Figure 5.1.

With responsibilities for maintaining quality systems standards passing to ISO, the need to keep the BS EN ISO 9000 series up to date, already mentioned, has to be considered in a wider than UK context. Within ISO there is a view that, as the ISO 9000 series is quite new to many countries, there should be a period of consolidation before any radical revision. In the meantime, a number of less radical changes have been suggested, mainly to remove ambiguities and ensure a degree of simplification and rationalization, such as the following:

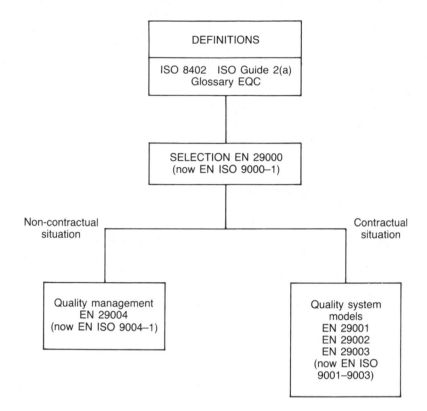

Figure 5.1 Relationship between international and European quality systems standards.

- Align the text in ISO 9001 and ISO 9002 for management responsibility, internal audits, and training.
- Take into account servicing and maintenance of equipment.
- Clarify requirements for quality plans.
- Provide sufficient and appropriate guidance on the use of statistical techniques.

More radical changes are postponed until 1999 or even later, when, with more countries having experience in the use of the ISO 9000 series, issues then requiring attention are likely to include:

- responsibility of personnel for safety;
- disposal of unwanted or unintentional products arising from processes;
- management of change and quality improvement;
- education methods and training;
- the marketing and sales processes; and
- the special needs of building and construction.

FURTHER READING

Much has been written about quality assurance and quality system standards. An early review on quality assurance and associated certification schemes is G. Atkinson, *A Guide Through Construction Quality Standards* (Van Nostrand Reinhold (UK), 1987). Other publications include: *A brief guide to BS 5750 for design organizations* (BSI Quality Assurance, PO Box 375, Milton Keynes MK14 6LL); BSI Handbook 22: Part 1 *Quality assurance*, Part 2 *Reliability and maintainability* (1992), which collects together current British and ISO Standards, including drafts for development; and BS 4778: *Quality terminology, Part 1: 1987* International terms, which is identical with ISO 8402: 1986 *Quality vocabulary*.

A number of organizations have issued publications, including: Cement and Concrete Association, now British Cement Association, *Quality Management Guidance Notes: Fundamentals of quality assurance on site* (1986), and *Planning for quality on site* (1989); CIOB, *Quality assurance in building*, 2nd edn (1989), and *Quality assurance in the building process* (1989); CIB, *Quality Assurance in Building – Working Papers*, Interim Report on work of CIB W 88 – Quality Assurance (CIB Publication 109, 1989); and A. Foster, *The application of BS5750: Quality systems to building services* (BSRIA, 1984).

CIRIA publications on quality assurance include:

SP 49: *Quality assurance in construction – the present position* (1987)
SP 50: *National Quality Assurance Forum for Construction*, Proceedings of conference (June 1987)
SP 55: *A client's guide to quality assurance* (1988)
SP 72: *Quality management in construction – certification of product quality and product management systems* (1989)
SP 74: *Quality management in construction – interpretations of BS 5750:1987 – quality systems for the construction industry* (1990)
SP 84: *Quality management in construction – contractual aspects* (1992)
SP 88: *Quality management in construction – implementation in design services organization* (J.N. Barber, 1992)
SP 89: *The impact of European Communities' policy on quality management in construction* (1994)
Report 109: *Quality assurance in civil engineering* (1985)
TN121: *Sample Quality Assurance Documents* (1985)

There have been two international reports on aspects of quality assurance. The application of quality assurance to consulting engineering firms in 12 European countries is the subject of a 1990 CEBI/CEDIC Task Force on Quality Assurance: *Status report on quality assurance in European countries consulting engineering services* (CEBI/CEDIC, avenue Louise 430, BTE 12, B-11050 Bruxelles). Secondly, the International Union of Building Centres published in

1990 *Directory of quality approval systems and assessment procedures* (UICB, PO Box 20754, 3001 JB Rotterdam).

There are also a number of books, including: B. Dale and J. Oakland, *Quality Improvement Through Standards* (1994); J. Duncan, B. Thorpe and P. Sumner, *Quality Assurance in Construction* (Gower Publishing Company, 1990); M.J. Fox, *Quality Assurance Management*; and T. Hughes and T. Williams, *Quality Assurance: A Framework to Build On* (BSP Professional Books, 1991).

Fitness for intended use | 6

Fitness for purpose: A construction product or work is fit for its purpose, if it can be used according to its destination for a defined purpose under fixed conditions taking into account the economic aspects.

Durability: A construction product or work is durable if it is designed and constructed in such a way that under conditions of normal maintenance and use the essential requirements are satisfied during an economically reasonable working life.

German proposal for EC SCC Guidance Paper

Unless there has been a long period of relevant service experience assessment of durability will have to be made on evidence from: (a) service experience of the material or component under the required conditions, but for a limited period – possibly under an Agrément Certificate.

PD 6501: Part 1: 1983 E.2 *Assurance of durability*

6.1 THE *AGRÉMENT* CONCEPT

To assess the fitness of a material, product or structure for an identified use has been the task of architects, engineers and other constructors throughout the centuries. Few have done it systematically. In the 1930s, the Building Research Station developed a set of functional criteria against which to assess systems of flat construction. The criteria were elaborated in 1942:

> to investigate alternative systems of house construction used in inter-war years, and to advise on such methods as may be capable of application or suitable for development in the post-war period.

In the early 1950s, the criteria were used to assess innovatory, and some less innovatory, proposals for non-traditional house construction. However, systematic assessment of building materials and techniques was abandoned when private house-building was resumed.

In France, to meet a similar need, a system was introduced for the technical assessment and acceptance of new materials and building methods for state-aided housing. It was called *agrément*. P. Roger, who played a major role at CSTB in its development, told the 1962 CIB Congress 'Innovation in building' that

> no *agrément* has any sense unless specific use(s) are precisely indicated ... An *agrément* is not a specification, but a technical appraisal to be used by technicians. Knowledgeable clients insist that new materials used in their buildings should be the subject of an *agrément*.

French *agrément*, under the then newly established CSTB (Centre Scientifique et Technique du Bâtiment), was to flourish for two reasons:

- the French civil code placed a more clearly defined responsibility of building damage during the first ten years on designers and builders;
- French decennial damage insurers agreed to accept a product given an *agrément* certificate, after review by their technical committee, as being no more of a risk to insure than a traditional product.

In *L'Assurance Construction* (1988), Pierre Maurin, secretary-general, l'Association Française des Assureurs Construction, has set out the French approach:

> Use of an innovatory process carries unknown risks, as does the use of new materials ... to deal with the problem the Association has in its standard decennial responsibility insurance contract a reference to *les ouvrages à risque normal*, accepting as a normal risk: (a) traditional products and techniques covered by French standards or the rules and specifications in DTUs; and (b) for non-traditional products and techniques an *Avis technique* (successor to French *agrément*) issued through CSTB, after examination and acceptance by the Association's technical committee.

6.2 AGRÉMENT IN THE UK

In the UK, there was no similar decennial responsibility legislation or systematic acceptance of technical risk by insurers. The 1950s saw designers specifying many materials for use in construction that have since proved to have properties and a performance that were inadequate for the use to which they were put, and not until the mid-1960s was any interest taken here in the *agrément* concept despite a BRS study.

When there was interest, the initiative came from architects, not from the insurance industry or manufacturers. In autumn 1963, Gerard Blachère, the CSTB director, was invited to talk about the French experience at a special RIBA meeting. After the meeting, Geoffrey Rippon, the then Construction Minister, accepted a recommendation from architects and others to give the Ministry's support to the

introduction of a system akin to French *agrément*. An independent
Agrément Board, appointed by the Minister, was set up; and in 1966
Dr T.W. Parker, its first director, opened an office at Hemel
Hempstead. Initially BRS acted as the Board's technical agent. Three
years later it became directly responsible for the assessment and tests
before an Agrément Certificate was issued. Since then, the Board has
issued over 2500 certificates.

In France, as a result of pressure for greater involvement in the
assessment process by manufacturers, overall responsibility was transferred to a government committee on which industry, insurers
and other interest are represented. *Agrément* became *Avis technique*,
to indicate that it was an assessment of the fitness of products rather
than an approval. CSTB continued to manage the system, being
responsible for most of the technical work. The new *Avis techniques*
continued to be accepted by decennial damage insurers, after review
and possibly additional explanations by their technical committee.

In the UK, where damage liabilities, being less clearly defined,
depend more on legal than technical considerations, the insurance
industry has, up to now, paid little or no attention to Agrément
Certificates as an instrument in risk assessment. However, they are
one of the technical supports to building regulations through Approved
Document Regulation 7: *Materials and workmanship*. From time to
time, the British Board of Agrément issues, through HMSO, a *BBA
Supplement to the Approved Documents Supporting the Building
Regulations 1985 and 1991*, in which appropriate Agrément Certificates
are listed.

In 1983, to distinguish between *agrément* and compliance with
requirements of a published British Standard, the respective responsibilities of BBA and BSI were agreed in a statement of mutual understanding, under which BBA offered to assess materials, products,
systems and techniques for Agrément Certificates when there is no
relevant British Standard or, if a relevant British Standard exists, when
the subject of the certificate either contains some innovative element
of manufacture or use and has a level of performance in use at least
as good as that of the material, product, system or technique complying with the British Standard, or has a performance in use that is
substantially better than that of the material, product, system or technique complying with the British Standard.

In the UK, the DoE has designated the British Board of Agrément
as a body authorized under 89/106/EEC Article 10 to issue European
technical approvals. BBA, as the single UK-designated body, is one of
the spokesbodies on the European Organization for Technical
Approvals (EOTA), its director and technical staff taking a leading
role in the development of EOTA operational procedures and ETA
guidelines. In most other Member States, as Box 6.1 shows, only one
ETA body has been designated.

BOX 6.1

European Organization for Technical Approvals:
Bodies designated by Member States to issue ETAs #

(from time to time revised lists of names are published in the Official
Journal of the European Communities: C series)

Belgium	– DGV Directie Goedkeuring en Voorschriften DAS Direction 'agrément et spécification'
Denmark	– ETA-Danmark A/S
France	– CSTB Centre Scientifique et Technique du Bâtiment
	– SETRA Service d'Études Techniques des Routes et Autoroutes
Germany	– DIBt Deutsches Institut für Bautechnik
Greece	– ELOT Hellenic Organisation for Standardisation
Ireland	– IAB Irish Agrément Board
Italy	– Servizio Technico Centrale della Presidenzia del Consiglio Superiore dei Lavori Pubblici
	– Centro Studi ed Esperenze Anticendi dei Corpo Nazionale dei Vigili dei Fuoco
	– ICITE Istituto Centrale per l'Industrializzione e la Tecnologie Edilizia
Luxembourg	– Laboratoire des Ponts et Chaussées
Netherlands	– SBK Stichting Bouwkwaliteit
	– BDA Keurings- en Certificeringsinstituut BV (roofing and insulation materials)
	– BMC Stichting Betonmortelcontrol (ready-mix concrete)
	– IKOB Stichting Instituut voor Bouwmaterialen (ceramic materials and elements)
	– INTRON Instituut voor Materiaal- en Milieu-onderzoek (concrete repair process)
	– KIWA Keuringsinstituut voor Waterleidungartikelen
	– Stichting Kwaliteitscentrum Gevelelementen (facade elements)
	– SKH Stichting Keuringsbureau Hout (timber products)
Portugal	– LNEC Laboratorio Nacional de Engenharia Civil
Spain	– IETCC Instituto Eduardo Torroja de Ciencas de la Construccion
United Kingdom	– BBA The British Board of Agrément

The address of EOTA is given in Part 2.1; that of a body authorized to issue
European Technical Approvals under the relevant national section in Part 2.2

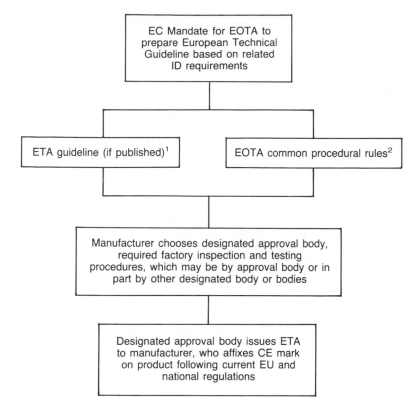

Figure 6.1 European technical approval (ETA) route to CE mark.

[1] Guidelines will incorporate product classes and attestation of conformity system given in Commission mandate, and indicate the appropriate methods of testing and assessment.
[2] These require, in the absence of an appropriate guideline, reference to appropriate Essential Requirements and Interpretative Documents. ETA bodies will submit to each other their draft ETAs for consultation to ensure compatibility.

6.3 EUROPEAN TECHNICAL APPROVALS

For the *agrément* concept, the most important event has been the adoption of European technical approvals as one of the two routes to CE marking under 89/106/EEC, as shown in Figure 6.1.

89/106/EEC describes a European technical approval as

> a favourable technical assessment of the fitness for use of a product for an intended use, based on fulfilment of the essential requirements for building works for which the product is used.

Conditions under which the ETA route is followed differ little from the Agrément route: for example, ETAs cannot be issued where there is a harmonized European standard, and like Agrément an ETA has a limited validity, usually five years but renewable.

As Figure 6.1 shows, ETAs may be based on guidelines prepared by EOTA under a Commission mandate. Guidelines are the documented assessment methods that underlie ETAs. Their general format, which has been agreed by EOTA, sets out the three major steps in the assessment process:

- translation of the essential requirements to requirements directly applicable to the relevant uses of the family of products being considered;
- listing of available verification tools: tests, calculation methods, etc.;
- bringing these items together in an appropriate technical specification.

Progress in the development and operation of ETA Guidelines is reported in the EOTA bulletin and in *BBA Monthly Datafile*.

In a chapter devoted to 'fitness for intended use', it is appropriate to examine two associated matters: durability and serviceability, and environmental protection. Both, in quite different ways, affect building quality.

6.4 DESIGN LIVES AND SERVICEABILITY

Most buildings are expected to have a long life. Many components and installations are expected to give, and actually achieve, good service indefinitely. But it may be neither practical nor economic to expect all components, and particularly all items of building services equipment, to last as long as the basic structure. Yet there has been only limited attention given to such issues as what is, or what is meant by, 'economic working life', or how to plan for replacement of products and installations that have reached the end of their useful working lives. Only to a restricted degree do the Building Regulations deal with the matter, although there is an interesting approach in the French Civil Code, which distinguishes between components fixed to a structure – *gros ouvrages* – and those that are more easily replaced – *menus ouvrages*.

If such an approach become general, it might be possible to accept a certain percentage of buildings as having design lives of a given number of years, thus making it possible to legislate for special and innovative construction in a way that will make the discussion of durability meaningful.

On longevity and adaptability the 1978 BRE report *A survey of quality and value in building* stressed the fact that half the dwellings in use in Britain a hundred years ago, and about 90% of those in use 50 years ago, are still in use today. It suggested that the effective life of a house in Britain was certainly over 50, and probably over 100 years, at least for the primary structure.

The 1983 BSRIA study *Quality assurance in building services* discusses in some detail system reliability, suggesting that manufacturers of building services equipment might consider the following:

- What life are their products expected to have?
- Do they design their products to have a defined life?
- Is life expectancy information given to users?
- Are records kept to check whether expected life is achieved?

Today, arrangements for disposal of equipment, building components etc. that are no longer usable have become a matter of importance in any quality programme, for a number of reasons:

- a marked increase in the volume of disposal items, a result of changes in both consumer needs and fashion and technical developments;
- the greater use of non-traditional materials such as plastics, lightweight alloys, and chemically treated wood derivatives, some of which require careful handling, storage and disposal because they contain radioactive, toxic or otherwise harmful materials, or are not easily biodegradable; and
- as a corollary, the stricter requirements of environmental protection legislation and enforcement, to which reference in made in the following section.

Despite the growing importance of safe disposal, there is no reference to this 'green' issue as being considered in the EC mandates to CEN. Unless arrangements for disposal at the end of a construction product's useful and economic life are given a place in harmonized European standards, it may later become a matter for separate national or European legislation and thus adding to an already complex situation.

6.5 ENVIRONMENTAL QUALITY

The safe use of sites, and control of pollution of land and watercourses, is an essential component of construction quality. To understand the issues involved is important for public authorities and commercial interests, and their advisers. But it is only a first step. Although yet rarely practised, it should be a duty of a local authority to maintain appropriate site records and ensure that developers deposit with them the kind of information that may be needed when a works or installation reaches the end of its useful, economic life and must be demolished and material safely disposed of. Moreover, such information should be in the public domain to facilitate changes of ownership or any major redevelopment. Without such information, there is growing evidence that the sale of sites and buildings becomes difficult and insurance against environmental risks costly, if not impossible.

Increasing attention is being directed at the environmental quality of a firm's activities, and – a closely associated topic – how it manages potential risks to the health and safety of employees, and more generally to the community. Here, BS 7750: 1992 *Specification for environmental management systems*, the world's first standard of its kind, gives guidance. It aims to get firms to think about the environmental aspects of their activities, acting as a checklist for preparation of publicly available environmental policy documents in which environmental targets are specified subject to regular internal audits.

Modelled on the quality system approach set out in the BS 5750 series, it provides a framework for organizations to manage environmental performance that can be audited and is capable of certification in parallel with or separate from BS 5750. It is expected that a European standard on environmental management systems, developed to support the European Commission's proposal for an eco-audit regulation, will be based on BS 7750.

To support this strategy, BSI is working through representation on CEN and in the ISO/IEC Strategic Advisory Group on the Environment (SAGE). In the meantime, in anticipation of the proposed EU regulation, the BSI technical committee responsible for BS 7750 has been working with a number of organizations in a pilot programme to assess the new standard in use.

An environmental audit is a first step in the management of an enterprise's environmental performance, in part to avoid costly eco-litigation and in part to meet the requirements of increasingly tough environmental regulations. The types of audit include:

- site audits directed at materials used, waste management, operation of anti-pollution measures;
- compliance audits directed at ensuring compliance with all relevant environmental regulations;
- investment audits directed at environmental implications of new capital investments in plant etc.;
- takeover audits directed at planned acquisitions to assess possible damage liabilities;
- activity audits directed at activities that cross conventional business boundaries, such as distribution of potential toxic materials;
- issue audits directed at a particular matter that is the subject of general concern, such as ozone-depleting chlorofluorocarbons;
- associate audits directed at a firm's suppliers;
- full corporate audits directed at the enterprise's activities as a whole from an environmental standpoint.

In many cases, especially where environmentally sensitive issues are involved, a first comprehensive audit will be followed by yearly checks. Depending on the activity and the size of the enterprise, an audit may be undertaken within the business, by a special unit within the enterprise or by a third party. A number of consultants, including environmental advisory units at universities, polytechnics and the BRE, are offering audit services.

FURTHER READING

For an account of the development of *agrément*, and details of the 1983 BBA/BSI memorandum of Agreement, see pp. 73–82 in G. Atkinson, *A Guide Through Construction Quality Standards* (Van Nostrand Reinhold (UK), 1987). The European Union of Agrément's periodical: *UEAtc information*, in English, French and German, reports on progress in the Union and at EOTA, as well as in member institutes. *BBA Monthly Datafile* carries news of the British Board of Agrément. It is supplemented by a quarterly *Index of Current BBA Publications*.

Progress in the development of EOTA policy and procedures, and in the mandating of ETA Guidelines, is reports in a bulletin published by EOTA, rue du Trône 12, B-1050 Brussels.

In January 1994, the Commission issued a Decision on 'common procedural rules for European technical approvals' (OJEC 20.1.94).

The Single European Market

7

Despite its name – a Common Market – when the barriers are down it does not mean a uniform market – regional and personal tastes will persist. There may be fewer officials barring our way, but you will still have to convince the buyer that your product or service is what he or she wants.

Introduction, *Gateway to Europe*, NEDO (1990)

7.1 THE SINGLE MARKET AND THE NEW APPROACH DIRECTIVES

Despite the elimination of tariff and quota restrictions between Member States envisaged in the Treaty of Rome, free movement within the European Union has continued to be impeded by barriers such as technical legislation, and the national standards used as back-up. In a first attempt to overcome this problem, a series of 'Old Approach' directives were adopted in which specific requirements, identical across the Union, were laid down for single products, based where possible on Europe-wide standards derived from national standards. It was a slow, continuing process as Table 7.1 demonstrates.

During the 1980s the Community faced a series of problems from which no relief seemed available until, after much debate, the 'New Approach' was given substance in the 1985 Single European Act – not ratified by all Member States until July 1987 – under which heads of government committed their governments to completing the internal market by 1 January 1993. To speed up decision-making, the Single European Act introduced qualified majority voting for all but a few key matters. There was agreement on a seven-year timetable for the removal of some 300 barriers to trade, which led to a 'New Approach to Technical Harmonization and Standards'.

In 1988 Lord Cockfield, a Commission vice-president in the 1980s and the driving force behind the New Approach, wrote:

> We are moving inexorably towards economic union in Europe. As a former businessman and United Kingdom, I am very concerned that British manufacturing and services industries gear up to respond effectively to this great challenge as their counterparts are doing across the Continent . . . The time to act is now. The creation

Table 7.1 Four steps towards a Single Market

	No internal tariffs	Common external tariff	Free movement capital and labour	Common economic policy
Free Trade Area (EFTA)	Yes	No	No	No
Customs Union (Benelux)	Yes	Yes	No	No
Common Market (EC)[a]	Yes	Yes	Yes	No
Economic Union (EU)[b]	Yes	Yes	Yes	Yes

[a] European Community after 1986 Single European Act
[b] European Union post-Maastricht Treaty

of a single integrated European economy is neither a revolutionary nor a vain aspiration. It was the original intention of the Treaty of Rome which established the Community 30 years ago – and should have been completed long ago. In the lean years of the 1970s and early 1980s the momentum was lost. But now it has been found again – with a vengeance. You can feel this wherever you go in the European Community. We are now well on the road to achieving European union in the economic field.

Somewhat optimistically he added:

The entry into force of the Single European Act provides us with the institutional means for making this plan a reality. It defines the internal market as 'an area without internal frontiers'. Note the words 'without internal frontiers' because we have to come back to them time and time and time again. The definition of the Single European Act then continues '... in which the free movement of goods, persons, services and capital is ensured ...' and it says this is to be achieved '... over a period expiring on 31 December 1992'.

Against the European Court's *Cassis de Dijon* decision that any product legally placed on the market in one Member State had to be allowed onto the market of all Member States, the New Approach was based on the following principles:

• Directives were to be applied only to areas covered by existing national laws or regulations.
• Legislation was usually to refer to 'harmonized' European standards.
• Harmonized European standards were in general to be written in performance and not prescriptive terms.
• Standards might specify different levels of performance depending on intended end use.

- Those responsible for national regulations could select levels or classes of performance within a harmonized European standard, related to local conditions and existing regulatory requirements.
- A CE mark, placed on a product by a manufacturer, would demonstrate to officers responsible for enforcement of national regulations that one or more essential requirements were satisfied in a specified way, and therefore the product could be placed on the market.
- The fact that a product carries a CE mark only signifies that, if used correctly in construction works, its use satisfies relevant regulations.
- The presence of a CE mark on a product does nothing more; it does not indicate that the product is of a higher level of performance than that required in regulations, or satisfies requirements set out in a non-harmonized standard or, for example, in a client, designer or manufacturer specification outside the regulatory system.
- Governments and public bodies cannot, without very good reasons, specify for works or supply of services, covered by a 'procurement' directive, a performance related to essential requirements that goes beyond that expressed in an identifiable grade or level in relevant harmonized European standards.

Thus emphasis was placed on the preparation, and adoption within the Union, of New Approach European standards: to underpin technically the New Approach directives; and to be the basis of technical specifications through which all who compete for public contracts do so on level terms.

Usually the standards would be in two parts: a harmonized part related to one or more 'essential requirements' and, therefore, to health and safety requirements of national regulations; and, a non-harmonized part, which might cover non-regulatory, usually commercial, requirements. The cost of preparing the first would be borne by the Commission on behalf of the Union; that of the second part would fall on European industry.

As testing, inspection and certification procedures could equally give rise to technical barriers, they would also be harmonized, although it was accepted that the level of conformity attestation in a standard could differ depending on end use.

To assist UK industry under the New Approach to a Single European Market, Department of Trade and Industry (DTI) has issued a number of explanatory booklets. A briefing pack summarizes the guidance. Late in the day, the Commission plans to issue a *New Approach Guide*, which aims at being an authoritative overview of key New Approach features, such as the meaning of 'placing on the market'. Special attention is given to the transitional period, i.e. the period between the date on which a directive enters into force and a subsequent date, set in each directive, when it is fully implemented in national legislation. Other matters dealt with in some detail include:

- conformity assessment procedures;
- responsibilities of Member States in connection with notified bodies designated to carry out these procedures;
- market surveillance by public authorities; and
- the role and scope of mutual recognition agreements.

7.2 THE NATURE OF EUROPEAN UNION LEGISLATION

> Scots law was, and is, like the Romano-Germanic systems, a unitary system.
>
> *The Legal System of Scotland*, HMSO, Edinburgh (1977)

An aspect of the creation of a Single European Market that may be foreign to the English, but not to the Scots, is that the legislation resulting from European Union decision-making procedures is based on Romano-German systems, and not the hybrid of common law and equity that is so rewarding for the English legal profession. European legislation, which is changing, or at least having a significant influence on, the national legislation of Member States has, as a model, continental law rather than the wordier, less user-friendly law of England and Wales.

The form taken by a directive is very similar to that of a French *décret*, as will be seen if the reader compares the following extract from the *décret* that transposed the Supplies Directive into French law with the English version of 89/106/EEC, the Construction Products Directive.

> Journal Officiel (French Republic)
> Décret No. 89-236 du 17 Avril 1989
> Modification du Code des marchés publics (transposition de la directive 'fournitures' en droit français)
> Le prime ministre, sur le rapport du ministre . . .'

(There follows a list of 11 ministries consulted.)

> Vu le Code des marchés publics
> (the collection of legal instruments relating to public contracts);
>
> Vu le décret . . . aux marchés des collectivités locales et des leurs établissements publics . . .'
> (the decree that extended the public contract code to local authorities and other public agencies);
>
> Vu le traité . . .
> (the Treaty of Rome);
>
> Vu l'avis de la commission centrale des marchés . . .
> (the opinion of the central committee on public contracts);
>
> Le Conseil d'Etat (section des finances) entendu . . .

(The State Constitutional Council has examined the decree and found it does not conflict with the constitution and any related laws.)

Décrète:

(There follow three short Articles. Art 1 inserts in the Code a new chapter, Livre V, on public contract procedures required by the EC directive; Art 2 cancels earlier decrees on public contracts; Art 3 gives the date on which the decree comes into force.)

As an Annex to the Decree, the eleven Articles of Livre V are cited:

Dispositions particulières applicables à certaines marchés de fournitures
(special procedures applicable to certain public supplies contracts.)

For reference by the French construction industry, the whole document had been reprinted on an A4 page in a *Textes Officiels* supplement in the weekly construction journal *Le Moniteur*.

As is the practice in France, and most other continental countries, where decrees and regulations are published in an official journal, directives and other legal instruments of the Union are published in the *Official Journal of the European Union*. Once adopted by the Council of Ministers, directives are binding on each Member State. How the transposition is made, and the choice of form and method of publication, however, are left to national authorities. It may be, as in the case of France's transposition of one of the public procurement directives by a short décret, by inserting a short additional chapter in an existing Code on Public Contracts, or, as in the UK, in a Statutory Instrument, such as the 1991 Construction Products Regulations, made under powers conferred by Section 2(2) of the European Communities Act 1972.

To ensure that Member States implement a directive, there is, at the end, a provisions chapter giving a date by which Member States shall 'bring into force the laws, regulations and administrative provisions necessary to comply with the directive'. Member States are usually required to send to the Commission copies of the legal texts through which the provisions of the directive are implemented.

A number of factors affect the speed with which Member States implement Union legislation, and the form of the resulting legal instrument:

• how well a Member State has consulted national interests during the proposal to adoption stages, how efficient are parliamentary procedures and national bureaucracies. The UK has a good record on consultation with industry and other national interests, but still depends on how well representative associations and professional institutes inform and consult their members;

- whether the central government can legislate for a matter like construction or, as in Federal Germany, it is for Länder legislation; and
- national legal traditions (here the UK suffers compared with those Member States with their codified Roman-based legal systems).

7.3 TYPES OF EUROPEAN LEGISLATION

The European Union decision-making procedures result in legislation that changes, or at least will have a significant influence on, the national legislation of Member States and its administration. Legislation is of five kinds: directives; regulations; decisions; recommendations and opinions; and case law.

7.3.1 Directives

Widely known because of their technical as well legal implications are Council directives. Originating in a Commission proposal, a directive is adopted by the Council of Ministers in cooperation with the European Parliament, usually after receiving the opinion of the Union's Economic and Social Council. Article 189(3) of the Treaty of Rome states:

> A directive shall be binding, as to the result achieved, upon each Member State to which it is directed, but shall leave to national authorities the choice of form and methods.

Council directives, which normally are the main form of European legislation, influence construction quality in a number of ways, such as determining:

- the quality and safety of products used in construction works;
- the qualifications and conditions of employment and establishment of construction professionals;
- the procedures for tendering for and awarding contracts for public works and utilities, and supply of goods and services;
- health and safety on construction sites, and in other workplaces;
- the use of hazardous substances and processes;
- matters concerned with environmental protection.

Draft directives, which are proposals of the Commission, have to undergo the process of informal and formal consultation and debate before being adopted by the Council of Member State Ministers, a process in which in the UK the Department of the Environment plays a major role where construction is the topic.

The speed with which Member States implement directives through national legislation depends in part on how efficient is a national bureaucracy and the country's parliamentary process. Some Member States, such as Italy and Greece, are slow to implement directives. In matters that are a *Länder* rather than a Federal responsibility, like construction, Federal Germany has faced problems in implementing some directives. France should have no administrative difficulty, but can take its time where the legislation affects one or another vested interest. Denmark, the Netherlands and the UK, in contrast, have a good record of implementation, due in large part to well-established consultative arrangements with industry and the construction professions.

7.3.2 Regulations

Regulations, the second type of legal instrument, are made under the authority of Article 189(3) of the Treaty of Rome:

> In order to carry out their task the Council and the Commission shall make regulations ... A regulation shall have general application. It shall be binding in its entirety and directly applicable in all Member States.

Like Council directives and Commission or Council decisions, regulations are published in the *Official Journal* L Series, and, in the UK, fall within Section 2(1) of the European Communities Act 1972 and require no further legal enactment.

7.3.3 Decisions

Through authority derived from the Treaty of Rome, the Council or Commission, directly or on the basis of a regulation or directive, may address a binding decision to a government, enterprise or individual. Article 189(3) of the Treaty states that:

> In order to carry out their task the Council and the Commission shall (inter alia) take decisions ... A decision shall be binding in its entirety upon those to whom it is addressed. It may be addressed to Member States or individuals and shall take effect upon such notification. A Member State would, if necessary, rely on national legislation to bring EC decisions into direct effect.

7.3.4 Recommendations and opinions

There are three other ways in which Union institutions influence national and industrial policies and practices. The first two – recommendations and opinions issued by the Commission or a directorate-general, the Economic and Social Council or the European Parliament – are not binding but may be useful for the promotion of government or sectoral business at a European level, an example being the series of Guidance Papers issued by the Commission's Standing Committee for Construction. Although the Commission has stressed that these Papers do not attempt to be a legal interpretation of the Construction Products Directive, are not judicially binding and cannot modify or amend the Directive in any way, they are useful, being 'primarily of those giving effect to the Construction Products Directive from a legal, technical and administrative standpoint'.

7.3.5 Case law

Lastly, national legislation may be modified through European case law, the result of decisions taken by the European Court of Justice in its task of interpreting directives and other Union legal instruments, as well as adjudicating between Union institutions or between any of them and one or more Member States. In these areas the Court's decisions overrule decisions made by national courts of Member States.

7.4 UPDATING EUROPEAN LEGISLATION

The adoption of European legislation, and its implementation by Member States, issue by European institutions of legal and guidance documents, amendments to national laws and regulations etc. are continuing processes.

In the UK, a principal source of information on the progress of construction legislation is *DoE Construction Monitor*, incorporating *Euronews Construction*, published at approximately monthly intervals by *Building* magazine for the Department of the Environment's Construction Policy Directorate. Some Member State construction ministries issue similar updating series. The authoritative sources of information are the C (Official Information and Notices) and L (Legislation) series of the *Official Journal of the European Union*.

In assessing the impact of European legislation, it is useful to distinguish between the directives, regulations and other legal instruments issued by the European Commission, and any UK or other national legislation that implements EU legislation. It is the last that is legally binding on constructors and suppliers, subject possibly to interpretation by the European Court in Luxembourg. How European legislation is enforced depends on the manner in which it has been transformed into national regulations, and on the national authority made administratively or legally responsible for enforcement. For example, in the UK, enforcement in the case of the provisions of 89/10/EEC is by local trading standards officials, and in the case of regulations transposing the Health and Safety at Work Directives by Health and Safety Executive (HSE) officials.

7.5 TECHNICAL SPECIFICATIONS AND CE MARKS

A different kind of Euro-document is technical specifications such as ENs (European Standards), ETAs (European Technical Approvals) and HENs (Harmonized European Standards). To the layman the situation is confusing. Both ENs and HENs are prepared by CEN, but only HENs, prepared under a Commission mandate, serve as a principal route through which manufacturers and importers' agents are enabled to CE mark their products, and, with the second route (ETAs), are likely to be essential elements in tender documents for works and supplies covered by the European procurement legislation.

When issued by CEN, ENs and HENs replace national standards, and acquire the same legal status as the national technical specifications that they replaced, even though non-harmonized European standards cannot be used to support CE marks. The British Board of Agrément, and other members of the European Union of Agrément, are likely to continue to issue assessment certificates outside the scope of 89/106/EEC. Yet to be determined are cases where, additional to ETA requirements, other aspects of performance are assessed. There is a similar situation where a product is tested against a non-harmonized EN, and cannot bear a CE mark.

One component of Euro-legislation over which some uncertainty remains is the CE mark. Initially at least CE marking will largely concern manufacturers and importers' agents wanting to put products on the European market, and trading standards officials, who will have to police the use, or misuse, of CE marks. Not being quality marks, they may confuse rather than assist, designers and specifiers. No legally CE-marked construction product had been placed on the market by autumn 1994. When they appear, despite national legislation, there may well be uncertainty over their status compared with unmarked products already on the market for a number of years. In an attempt to reconcile somewhat different requirements in New Approach directives on CE marking, in June 1993 the Council of Ministers agreed a common position on formal proposals made by the Commission for a directive and Council decision on CE marking. The aim is to introduce a common set of rules across all directives requiring the affixing and use of CE marks. The Council decision will result in modifications to existing and new directives where there is reference to conformity assessment requirements. The amending directive is unlikely to come into force before 1 January 1995 at the earliest.

The European Economic Area came into force on 1 January 1994. It extended the application of most 'Single Market' measures. extended before 31 August 1991, to the EFTA countries.

FURTHER READING

For designers, constructors, regulators, manufacturers, suppliers and others concerned with construction, the DoE periodical *Construction Monitor* and, especially, its *Euronews Construction* is an updated source of information on European legislation, CEN activities and their impact on building and public works. To be included on the free mailing list, contact DoE, Room C1/02, 2 Marsham Street, London SW1P 3EB (tel: 0171 276 6698).

In the UK, the *Official Journal of the European Union:*

- C series (legislative proposals, resolutions passed by Council of Ministers, ESC opinions, and European Parliament Minutes)
- L Series (texts of approved legislation)
- S series (calls for tenders: public contracts, research contracts)

is available from HMSO, and from the EU London office.

BRE Digest 397, *Standardization in support of European legislation: what does it mean for the UK construction industry?* (November 1994), explains the reasons behind changes in the status of standards, and the processes involved in drawing up and implementing European standards supporting 89/106/EEC, the Construction Products Directive, and other European legislation.

Two reviews, the first written in 1970 and the second 18 years later, describe the ways in which Europe, and in particular the two founder members of the European Union – Federal Germany and France – achieve construction quality through standards and codes, and their application in building regulation and technical control, also the back-up provided by research, testing facilities etc.: G. Atkinson, 'The roles of authorities, designers and builders in Western Europe', 1970 Chartered Surveyors Annual Conference; and *Construction Research and Development in Federal Germany, France, Japan and the USA – an overview* (CIOB, January 1988).

There are many general publications on the European Community, and developments since Maastricht to create the European Union. They include:

R. Batley and G. Stoker (eds), *Local Government in Europe: Trends and Developments* (Macmillan Education Ltd, 1991);

The New Treaty on European Union (Belmont European Policy Centre, 42 boulevard Charlemagne, B-1040 Brussels, 1991);

Pocket Guide to the European Community (The Economist Publications, 1989);

P.S.R.F. Mathijsen, *A Guide to European Community Law* (Sweet & Maxwell, London, 1991).

Publications directed at construction in Europe include:

National Council of Building Materials Producers and *Building* Magazine, *The UK construction industry and the European community* (Building (Publishers) Limited, London E14 9RA, June 1988);

NEDC, *Gateway to Europe. A Manual for the Construction Industry* (National Economic Development Office, Millbank Tower, Millbank, London SW1P 4QX, March 1990);

F. Spencer Chapman and C. Grandjean, *The Construction Industry and the European Community* (Blackwell Scientific Publications, October 1991);

Engineering Consultancy in the European Community (Thomas Telford Ltd, London E14 4JD, 1990);

RIBA 1992 Committee, *Architectural practice in Europe: France; Germany; Italy; Portugal; Spain*, a series of four studies obtainable from RIBA Bookshop;

A. Burr, *European Construction Contracts*, a looseleaf series published by Wiley Chancery, Chichester.

The DTI *Single Market Business in Europe* series, issued free by the Department of Trade and Industry, aims to keep firms informed on developments within the European Union, particularly the implementation of European legislation. The following are some of the booklets available from DTI 1992, PO Box 1992, Cirencester GL7 1RN, or by telephone from 0181 200 1992:

Company law harmonization (February 1992)
Conformity assessment and what it means for business (April 1992)
Construction products – UK regulations (January 1992)
Electromagnetic Compatibility – the Directive in brief (May 1992)
Equipment for use in potentially explosive atmospheres (August 1993)
Financial services (March 1992)
Gas Appliances – the Directive in brief, 2nd edn (June 1992)
Guide to public purchasing, 4th edn (March 1992)
Guide to sources of advice (March 1992)
Keeping your product on the market – new common rules (April 1992)
Preventing new technical barriers (February 1993)
Public purchasing (March 1992)

DTI has also issued an Information Document: *CE marking. Your guide to the changes*, URN 94/634 (October 1994).

A number of European directories, some but not all published annually, are a useful source of reference. In London, most are available for reference at the City of London Business Library, 1 Brewers' Hall Garden, London EC2V 5BX (tel: 0171 638 8215).

The Construction Products Directive: 89/106/EEC **8**

For the purpose of this Directive, 'construction product' means any product which is produced for incorporation in a permanent manner in construction works. including both buildings and civil engineering works.

> 89/106/EEC Article 1.2

Member States shall not impede the free movement, placing on the market or use in their territory of products which satisfy the provision of this Directive.

Member States shall ensure that the use of such products, to the purpose for which they are intended, shall not be impeded by rules or conditions imposed by public bodies or public bodies acting as a public undertaking or acting as a public body on the basis of a monopoly position.

> 89/106/EEC Article 6.1

8.1 THE DIRECTIVE

The title of 89/106/EEC includes the phrase 'approximation of laws, regulations and administrative provisions'. It will mislead many. Beyond its preambles, 89/106/EEC makes limited reference to these provisions. The route taken to approximation is through the harmonization of technical specifications by means of which Member States may, or may not, satisfy regulatory requirements, and where they do may, or may not, make their use mandatory.

The aims of 89/106/EEC are twofold:

- to establish a single market in construction products by removing technical barriers to trade; and
- to prevent manufacturers placing products on the market that, when used in 'works' by competent persons, might result in an safe, unhealthy, environmentally unfriendly building or civil engineering work.

To ensure that products are 'allowed free movement and free use for their intended purpose throughout the Community', 89/106/EEC establishes conditions for a harmonized system of general rules for CE marking of construction products. These conditions lay down that products may be CE marked if they are fit for the intended use: that is, that they have to have such characteristics that the works in which they are to be incorporated or installed, 'if properly designed and built', satisfy essential requirements on 'safety and other aspects which are important for the general well-being', and are prescribed in national regulations or 'administrative provisions' of Member States.

89/106/EEC is probably the most complex of the New Approach European Union directives for construction practitioners, for the following reasons:

- Its scope extends to civil engineering as well as building works.
- It is centred on the specification of six essential requirements for works, mainly on matters to do with personal safety and health covered in most national regulations and 'technical specifications'.
- It is being implemented through:
    - six Interpretative Documents, prepared by expert groups and issued by the Commission, which serve as bridges between essential requirements for works and European standards for construction products;
    - the issue of mandates from the Commission to CEN, the European standards body on which EC and EFTA Member States are represented through their national standards organizations, to prepare harmonized European product standards (HENs);
    - for 'new' products, the establishment of a second European technical approval (ETA) route, managed by EOTA, a new European Organization for Technical Approvals, ETAs being issued by nationally approved ETA bodies, like the British Board of Agrément and France's CSTB, following guidelines mandated by the Commission and prepared by EOTA;
    - CE marking by manufacturers of products which either have an ETA or, depending on the requirement in a HEN, are certified by a nationally approved certification body, or declared by the manufacturer as meeting the standard's requirements.
- A CE mark is not a quality mark; it is a passport allowing the marked product free entry and use within the Union.
- It is accepted that different levels of performance, reflecting climatic and similar differences, may be set in a HEN, and public building owners may specify a level of performance appropriate to the needs of a particular project, but only with difficulty can they require any level higher than that set out in the HEN.
- Private building owners can, however, set any level they require as long as the level of performance is, at least, that required in national building and similar regulations.
- As well as HENs, compliance with which is mandatory and hence absolute, CEN technical committees may issue European standards

for matters not covered by essential requirements for health, safety etc., either as separate sections of the standard or as a separate document.

- There are three categories of ENs: category A – design, installation and execution standards, including, as prestandards (ENVs), eight structural Eurocodes; category B – product standards; and category Bh – standards for methods of test and measurement.
- When an EN has been published for a particular matter, national standards bodies are required to withdraw any of their standards covering the same matter within six months.

8.2 DELAYS IN BRINGING 89/106/EEC INTO EFFECTIVE OPERATION

Because of the complicated character of 89/106/EEC, and the possibility of overlapping directives, it has taken since December 1989 until mid-1993 to complete drafting of the six Interpretative Documents, which are the key to the preparation of harmonized European standards, and guidelines for European technical approvals.

Only in June 1993 did the Council of Ministers reach a 'common position' on an EU directive, and a decision on the affixing and use of CE marking. CEN technical committees have yet to finalize some 1000 ENs. And although, for example, the UK Construction Products Regulations 1991 came into force in December 1991 and gave, as a Statutory Instrument, powers to trading standards bodies over misuse of the CE mark, it is not yet effective.

In the short term, 89/106/EEC, whose objective was to remove barriers to trade in construction products, will do little or nothing to limit building damage and risk of latent building defects. In the longer term, possibly early next century, the general use of European product standards and European technical approvals, with associated arrangements for testing and certification, will have created a generally acceptable level of performance for products offered by manufacturers within the European economic area. Other measures, such as application of the EN ISO 9000 series of quality system standards to the whole process of regulation, design and construction management, the better training of personnel, and a better common presentation of technical information, may well make a more significant contribution to sound building.

8.3 THE ARRANGEMENT OF 89/106/EEC

The arrangement of 89/106/EEC is shown in Box 8.1.

Chapter I, in covering its field of application, defining key terms, and discussing essential requirements and their linkage to technical specifications, holds the key to the creation of a European market in construction products.

BOX 8.1 COUNCIL DIRECTIVE OF 21 DECEMBER 1988 ON THE APPROXIMATION OF LAWS, REGULATIONS AND ADMINISTRATIVE PROVISIONS OF THE MEMBER STATES RELATING TO CONSTRUCTION PRODUCTS (89/106/EEC)

Preambles in which the reasons for the Directive, and the procedures adopted because of the special nature of construction, are set out

CHAPTER I: Field of application – Definitions – Requirements – Technical specifications – Free movement of goods

CHAPTER II: Harmonized standards

CHAPTER III: European technical approval

CHAPTER IV: Interpretative documents

CHAPTER V: Attestation of conformity

CHAPTER VI: Special procedures

CHAPTER VII: Approved bodies

CHAPTER VIII: Standing Committee on Construction

CHAPTER IX: Safe clause

CHAPTER X: Final provisions

ANNEX I: ESSENTIAL REQUIREMENTS

ANNEX II: EUROPEAN TECHNICAL APPROVALS

ANNEX III: ATTESTATION OF CONFORMITY WITH TECHNICAL SPECIFICATIONS. 1 Methods of control of conformity; 2 Systems of control of conformity; 3 Bodies involved in the attestation of conformity: 4 EC Conformity Mark, EC Certificate of Conformity, EC Declaration of Conformity

ANNEX IV: APPROVAL OF TESTING LABORATORIES, INSPECTION BODIES AND CERTIFICATION BODIES

It is important to note that a construction product, as defined by 89/106/EEC, is anything 'produced for incorporation in a permanent manner' in either building or civil engineering works. A product suitable for construction works (as a whole and in their separate parts) must be fit for its intended use, account being taken of economy. However, 89/106/EEC only applies to products intended for use in works subject to regulations containing one or more of the following six essential requirements:

(1)   Mechanical resistance and stability

The construction works must be designed and built in such a way that the loadings that are liable to act on it during its construction and use will not lead to any of the following:

(a) collapse of the whole or part of the work
(b) major deformation to an unacceptable degree
(c) damage to other parts of the works or to fittings or installed equipment as a result of major deformation of the load-bearing construction
(d) damage by an event disproportionate to the original cause.

(2)   Safety in fire

The construction works must be designed and built in such a way that in the event of an outbreak of fire:

– the load-bearing capacity of the construction can be assumed for a specific period of time
– the generation and spread of fire and smoke within the works is limited
– the spread of fire to neighbouring construction works is limited
– occupants can leave the works or be rescued by other means
– the safety of rescue teams is taken into account.

(3)   Hygiene, health and the environment

The construction work must be designed and built in such a way that it will not be a threat to the hygiene or health of the occupants or neighbours, in particular as a result of any of the following:

– the giving off of toxic gases
– the presence of dangerous particles or gases in the air
– the emission of dangerous radiation
– pollution or poisoning of the water or soil
– faulty elimination of waste water, smoke, solid or liquid wastes
– the presence of damp in parts of the works or on surfaces within the works.

(4)   Fitness for use

The construction work must be designed and built in such a way that it does not present unacceptable risks of accidents in service or in operation such as slipping, falling collision, burns, electrocution, injury from explosion.

(5)   Protection against noise

The construction works must be designed and built in such a way that noise perceived by the occupants or people nearby is kept down to a level which will not threaten their health and will allow them to sleep, rest and work in satisfactory conditions.

(6) Energy economy and heat retention
The construction works and its heating, cooling and ventilation installations must be designed and built in such a way that the amount of energy required in use shall be low, having regard to the climatic conditions of the location and needs of the occupants.

It will be noted that these essential requirements apply to works, and only where the works are covered by regulations or, as 89/106/EEC describes them, 'national provisions laid down by law, regulation of administrative action'. Moreover, such requirements must, subject to normal maintenance, be satisfied for an economically reasonable working life, and only, in general, apply to actions that are foreseeable. They thus have a direct influence on the technical characteristics of products, incorporated in the works 'in a permanent manner', which, in turn, are reflected in national standards and other technical specifications, which may be a barrier to trade within the Union unless harmonized.

89/106/EEC's essential requirements differ little from requirements set out in the English Building Regulations. Despite differences in scope and arrangement, much of the other building legislation in the Union has similar requirements if the fact is accepted that there may be differences in climatic conditions, ways of life and degrees of protection that may prevail at national, regional or local level. It is a possibility that, in the words of Article 3.2, 'may give rise to the establishment of classes' in European technical specifications.

Article 4 explains that the term 'technical specification' includes harmonized standards, described in Article 4.2 and Chapter II, and European technical approvals, described in Chapter III and Annex II. It also refers to matters that, during the implementation of the Directive, have presented practical difficulties: recognition of national technical specifications (Article 4.3); minor part products (Article 4.5); and EC marking (Article 3.6), details being given in Annex III.4. Article 5 provides for the possibility that a Member State may wish to challenge the way in which essential requirements have been interpreted in technical specifications; also that the Commission, or a Member State, may consider a technical specification 'no longer fulfils the conditions necessary for the presumption of conformity'.

Article 6 is directed to Member States. It requires them not to impede the free movement, placing on the market or use of products that satisfy the Directive's provisions; and to ensure that any use for which a product is intended shall not be impeded by any rule or condition imposed by a public body, or private body acting as a public undertaking or as a public body on the basis of a monopoly position. It also requires Member States, during a transitional stage until European technical specifications provide otherwise, to allow products to be put on the market 'if they satisfy national provisions consistent with the Treaty'. A Member State can, however, prevent a product being placed on the market or, at least, impede its use where:

- the relevant European technical specification distinguishes 'between different classes corresponding to different performance levels', a possibility mentioned in Article 3.2;
- the Member State has determined the performance level to be observed in its territory;
- that performance level is not arbitrary but is within the classification adopted at Union level; and
- the product's performance is below that level.

The next three chapters – II Harmonized standards; III European technical approval, with Annex II, and IV Interpretative documents – describe the procedures to be adopted to bring into operation the essential instruments needed for the creation of an open European construction products market.

Chapter V, with Annex III, is devoted to attestation of conformity, detailing the two systems – declaration of conformity by manufacturer and certification of conformity by an approved body – which may be adopted in a technical specification on the basis of criteria given in Articles 13.4 and 5. Article 15, with Article 14.2 and Annex III.4, is devoted to the CE Conformity Mark, and associated CE Certificates and Declaration of Conformity.

Article 4.5 introduces the concept of products that play a minor part with respect to health and safety and in respect of which a declaration of compliance with the 'acknowledged rule of technology', issued by the manufacturer, will authorize such products to be placed on the market. The case of individual (and non-series) production, where a declaration of conformity will usually suffice, is covered in Article 13.5.

Chapter IX 'Safeguard clause' specifies measures available to a Member State concerned with shortcomings in the operation of the Directive. Chapter X 'Final provisions' sets two deadlines, now past: by which Member States brought into force the provisions of the Directive; and the date, 31 December 1993, by which the Commission, helped by the Standing Committee, 'shall re-examine the practicability' of the Directive procedures.

8.4 THE IMPORTANCE OF 89/106/EEC

89/106/EEC remains the most important for European construction of the New Approach directives adopted so far. And although, like other New Approach directives, its stated purpose is to establish a single market for construction products, the chosen route is unique for a number of reasons:

- Essential requirements, which have to be satisfied, are not for products but for the building and civil engineering works in which the products are 'installed in a permanent manner'.

- Interpretative Documents are used as 'necessary links' between these essential requirements for works and technical specifications for products.
- Technical specifications can be in the form of harmonized European standards, guidelines for European technical approvals, or, in some cases, ETAs on their own.
- CE marking can indicate either conformity to a relevant harmonized European standard, or the granting of a European technical approval by a body authorized by its Member State.
- Conformity to a harmonized European standard can be certified by an approved body, or by a manufacturer's declaration, depending on the nature and use of the product.

Its immediate importance is for suppliers because, in order to remove technical barriers to trade, it introduces new and possibly stricter conformity procedures; and, for constructors, because it aims to ensure that any product placed on the market has been 'guaranteed' by the supplier, or by an independent officially approved body, that it satisfies a technical specification based on an interpretation of the essential requirements for safety etc. of the building or civil engineering works in which it is intended to be used.

As explained earlier, the route taken is complex, and probably could not have been made less complex. Those responsible reasoned as follows:

- It is the way in which construction products are put together to become 'works' that is central to the regulatory process.
- For this reason, it is 'works' that are the subject of 'national provisions laid down by law, regulation or administrative action' (in French *dispositions legislatives, réglementaires et administratives*).
- These provisions exist in a wide range of national legal texts, and not just in building regulations but also obligatory specifications of highway and other public authorities.
- In general, 'national provisions' refer to technical specifications – standards and 'technical approvals' – which because of their disparity hinder trade within the EU.
- Therefore the route that the directive calls in the English version 'approximation' and in French *rapprochement* should be through the harmonization of technical specifications for products.

In theory, the approach adopted seemed sensible, and may well have been the only approach possible when it was decided to include all products 'produced for incorporation in building and civil engineering work' in a single directive. But to relate technical specifications for products to their fitness for safe use in works is a far from simple matter, and in practice the 'approximation' of national provisions has proved more difficult to achieve than the authors of the Directive anticipated when the Council adopted it at the end of 1988.

8.5 ESSENTIAL REQUIREMENTS

That the concept of requirements

> relating not only to building safety, but also to health, durability, energy economy, protection of the environment, aspects of economy, and other aspects important in the public interest

is central to the harmonization of technical specifications for construction products within the European Union has already been noted; as has the fact that national building regulations are often barriers to trade, even though they share common objectives – safety and health in buildings.

Of the six essential requirements, ER 1 is concerned with structural safety and damage limitation or, in the words of 89/106/EEC, 'Mechanical resistance and stability', a requirement usually given first place in national regulations, although its specification and the way compliance is ensured may differ. It was for this reason that the Commission's first approach to harmonization was to promote the drafting of the structural Eurocodes.

Fires in buildings cause greater human and financial losses than structural damage. ER 2: *Safety in case of fire* lists five separate groups of requirements. The first relates to structural fire performance, a long-established requirement in regulations, as is the third requirement limiting the spread of fire to neighbouring buildings. It is in the three requirements that relate to spread of fire, and especially the spread of smoke and toxic gases and their effects on the safety of occupants, firefighters and rescue teams, where there are the largest variations in the content and coverage of national regulations. 'Reaction to fire' testing has proved to be a stumbling block in the implementation of 89/106/EEC. Fire engineering of modern buildings is complicated by the variety of their design and construction, by the changing uses and occupancies to which they may be put, and by the hazards caused by smoke and toxic gases from burning materials, particularly polymers and composites. Relatively few Member States have updated regulations covering fire safety. Fewer still have adopted computer modelling as an aid to fire engineering.

Hygiene, protection of health, including protection against noise, and protection of the environment, are wide-ranging topics, some also being the subject of health and safety directives, the implementation of which is a responsibility (at least in the UK) of other than building authorities and which may apply to other than new works. Although this fact is unlikely to affect the development of harmonized European standards for construction products, there will be a need nationally to review the whole field of regulation if designers, builders and owners of premises are not to be 'over-regulated'. ER 6: *Energy economy and heat retention* covers matters already receiving much attention by Member States: for example, in the revisions to Approved Documents F and L of the English Building Regulations..

8.6 INTERPRETATIVE DOCUMENTS

The essential requirements shall be given concrete form in documents (interpretative documents) for the creation of the necessary links between the essential requirements . . . and the standardization mandates, mandates for guidelines for European technical approvals or the recognition of other technical specifications . . .

89/106/EEC Article 3.3

In themselves the six essential requirements are little more than a framework for listing the critical functions of building and civil engineering works. This framework, as stated in 89/106/EEC Article 3.3, has to be linked with the standardization tools required for the drafting of European technical specifications. These links are made through a set of Interpretative Documents, the preparation of which is a prerequisite to the issue of mandates to the European Standards Committee (CEN) for harmonized European standards, and to the European Organization for Technical Approvals (EOTA) for European Technical Approval guidelines.

Panels of experts from Member States and European organizations began work on the preparation of the Interpretative Documents at the end of 1989. Final drafts were not available for adoption by the Standing Committee on Construction until early summer 1993. They have now been published.

The route from essential requirements through Interpretative Documents, and Mandates from the Commission, to the issue of harmonized European standards and guidelines for European technical approvals must be completed before CE-marked products may be placed on the market. The delay in completing and publishing the Interpretative Documents has meant that, for example, the UK's 1991 Construction Products Regulations are unlikely to come into practical operation before some date in 1995. In the meantime, manufacturers have had to continued to supply products against national technical specifications or, where appropriate, an Agrément Certificate. None of these allows a product to bear the CE mark, and any supplier placing a CE-marked product on the market is liable to prosecution under the 1991 Regulations. In the absence of relevant technical specifications, manufacturers exporting to another Member State have had to continue to follow either existing arrangements, or adopt Article 16 procedures.

In a general introduction to the six Interpretative Documents, the Commission points out that, although the Interpretative Documents are exhaustive, they are of an evolutive nature. They are based on a combination of:

- the detailed development of the six essential requirements set out in 89/106/EEC Annex 1;
- knowledge of existing national building and other regulations applicable to works, including those in the field of procurement;

- the state of the art concerning construction products; and
- their intended use.

Any Interpretative Document is capable of amendment and elaboration in the light of technological development, and of progress in the state of the art concerning construction and construction products. In particular, at this stage, it was found necessary to limit the determination of classes and levels of performance to ID 2: *Safety in case of fire*. Otherwise, presumably, they will have to be determined by CEN and EOTA technical committees during the preparation of harmonized European standards and guidelines for European technical approvals.

One matter of wider interest is the statement that 89/106/EEC does not require a Member State to impose any one of the essential requirements in works: that is, Member States remain free to regulate or not regulate construction works. But, if they do, and their new or amended regulations have an impact on construction products, then the Member State must follow 89/106/EEC procedures, cannot impose any new essential requirement, and can only permit placing a product on the market if it is fit for its intended use.

Despite their evolutive nature, the Interpretative Documents can be regarded as authoritative summaries of acknowledged rules of technology applied to matters to do with safety, health and, to a lesser degree, conservation of energy, and they are prepared with the help of leading European practitioners, standards makers and research technologists, including those from the UK Building Research Establishment. They not only serve as guides for the preparation of harmonized European standards and guidelines for European technical approvals, but are also valuable general guides to the achievement of construction quality and will serve professional bodies and schools of architecture, building and engineering as frameworks for training manuals and educational textbooks. By making explicit essential requirements for safe and healthy buildings and civil engineering works and, through the Interpretative Documents, establishing a common terminology, those responsible for drafting 89/106/EEC have not only provided Europe with the means of creating an open market in construction products; they have also made available a framework around which a modern series of national building codes and regulations can be developed.

Each Interpretative Document is introduced by a part common to all, which explains the purpose and scope of Interpretative Documents, and deals with the meaning of seven 'general terms': 'construction works'; 'construction products'; 'normal maintenance'; 'intended use'; 'economically reasonable working life'; 'actions'; and 'performance'. The common part refers to levels or classes for essential requirements and related product performances; and to the basis for verification of the satisfaction of an essential requirement.

There is a general section on the treatment of the working life of construction works, and construction products, in relation to an essential requirement.

8.7 CLASSES IN TECHNICAL SPECIFICATIONS

That 89/106/EEC Article 3.2 introduced the concept of classes has been mentioned earlier. Buildings and civil engineering works are built in very different locations, have very different uses and are occupied by very different groups of people. They are subject to many different actions from climate and weather, and other environmental conditions. Risks from damage from actions by vehicles, from accidental or wilful actions of people, and from materials stored within or nearby depend on location and uses of works, as does the potential harm to occupants and passers-by. Different levels of performance and risk are accepted in different situations. Some, but by no means all, of these differences are recognized in national regulations and standards. Many affect the performance of, and therefore requirements from, construction products. Some reflect attitudes to risks to safety and health, ways of life, and uses of buildings, and characteristics of construction products, which may or may not still be valid in the light of modern technology, findings of scientific research or changes in social behaviour.

For these and other reasons different classes and levels of performance, and related procedures for attestation of conformity to technical specifications, need to be identified and made transparent, especially where their presentation and application in national regulations and standards are barriers to trade.

For an open market in construction products it is necessary:

1. to identify where classes are specified, the reason for their selection and their effects on design and construction;
2. to eliminate trivial distinctions; and
3. to get general agreement on classes that are important and on the specification of 'boundaries'.

This task falls, in the first place, to the Commission, on recommendations of the Standing Committee on Construction (aided by expert guidance), which has a responsibility for specifying different classes, levels of performance, and levels of attestation of conformity in standardization mandates to CEN and mandates for guidelines for European technical approvals. That the task is not easy is demonstrated by the difficulty that the Commission found in handling the matter of classes and levels of performance in the six Interpretative Documents.

Some idea of the kinds of classes that may have to be considered are:

- structural loadings, a consequence of effects such as seismic activity, climatic effects due to snow and wind, and man-made effects due to different risk of impacts from vehicles;
- the presence of abnormal levels of radioactive minerals, a result of local geological conditions;
- the risk of electrical storms;
- special protection against noise from aircraft and heavily trafficked highways;
- occupancy of buildings (most regulations distinguish between commercial, industrial and assembly buildings and dwellings; some between one-family houses and other types of housing, or dwellings only occupied as holiday homes).

For the present, therefore, it is being left to CEN/CENELEC technical committees to determine under mandate where specification of classes and levels of performance are necessary or superfluous, and to each Member State to choose from those adopted at European level. Where national levels above minimum requirements are adopted, these should be made transparent for the benefit of manufacturers, regulators and specifiers.

Once classes and levels of performance have been determined, constructors and regulators will have two tasks: to know what are the specific requirements of their national regulations, which may (but need not) be expressed in terms of European standard classes; and to check from the supplier's declaration that the product belongs either to that performance class or to a higher class.

During a transitional period, which may last some years, manufacturers may well find it no easy task to choose the best class in which to make and market a product; and some will not attempt to enter some markets. Regulators, possibly under pressure from industry, may well review the class or classes in their regulations, and bring them in line with standard European classes, as may public purchasing authorities.

8.8 IMPLEMENTATION OF 89/106/EEC

For the European construction industry and, especially, for the suppliers of construction products, 89/106/EEC is an important item of European legislation. When fully operational, it will have brought about the replacement of large parts of national product standards by new European standards more or less prepared to a common model. It will have seen European technical approvals for newer, or yet to be standardized, products supplementing or replacing national *agréments*. It will also have brought about new arrangements for approval by national governments of certification and inspection bodies, and testing laboratories; and, it is hoped, will ensure greater clarity in statements on the performance of construction products. It may or may

---

BOX 8.2 RESPONSIBILITIES OF MEMBER STATES UNDER
89/106/EEC

- To appoint two representatives to the Standing on Construction, who 'may be accompanied by experts'.

- To notify other Member States and the Commission of the names and addresses of the bodies the State has authorized to issue European Technical Approvals. 'The approved bodies designate by the Member States form an organization' (EOTA) to work closely with the Commission. 'Where a Member State has designated more than one approved body, the Member State shall be responsible for coordinating such bodies; it shall also designate the body which shall be the spokesman in the organization.'

- To communicate to the Commission, if a Member State wishes, the texts of national 'technical specifications', i.e. national standards and technical approvals, which a Member State 'regards as complying with the essential requirements'; and if notified by the Commission of 'presumption of conformity' publish references to the texts.

---

not have caused national governments to review existing systems of building regulation, simplifying requirements and procedures.

Already 89/106/EEC has brought about a greater degree of understanding of technical and procedural matters by representatives of national construction and industry ministries during more than 27 meetings of the Commission's Standing Committee on Construction, assisted in associated preparatory and working groups by experts from industry and the professions, research, standards, testing and other organizations. It has seen a reorganization and strengthening of the work of CEN/CENELEC relating to construction, in turn reflected in the UK in the structure of BSI, and in the relations between government, industry, the professions and other interests through the Department of the Environment's Joint Advisory Committee on Technical Specifications. It has brought into being a new European Organization for Technical Approvals (EOTA), and caused the European Union of Agrément (UEAtc) to examine its work and responsibilities.

Although implementing the directive through national legislation is one of the major tasks of Member States, they have other responsibilities under 89/106/EEC, such as the approval and supervision of two kinds of approved bodies: bodies involved in attestation of conformity to harmonized European standards; and bodies authorized to issue European technical approvals. These responsibilities are summarized in Box 8.2.

BOX 8.3 COMMISSION OF THE EUROPEAN UNION: DIREC-
TORATE GENERAL FOR INTERNAL MARKET AND INDUSTRIAL
AFFAIRS

- To arrange for the setting up of 'a Standing Committee on Construction' and to appoint a representative of the Commission as chairman.

- To instruct, after consulting the Standing Committee, technical Committees, 'in which the Member States participate', to prepare the Interpretative Documents giving 'concrete form' to six Essential Requirements relating to safety, health, protection against noise and energy economy.

- To arrange for the Interpretative Documents 'after soliciting the opinion of' the Standing Committee to be published in the EU *Official Journal*.

- After consulting the Standing Committee, to give Mandates to the European standards organizations (CEN/CENELEC) to 'ensure the quality harmonized standards . . . expressed as far as practicable in performance terms', and once they are established to have reference to them in the EU *Official Journal*.

One of the principal tasks of the Commission, now completed, has been to prepare the six Interpretative Documents. Box 8.3 lists this and other key tasks for the Commission.

The setting-up of a Technical Preparatory Group for Construction in September 1992 has done much to progress the implementation of 89/106/EEC. The six Interpretative Documents have reached their final form. EOTA has been set up, and work has started to agree procedures and the form of European technical approval guidelines. The distinction between technical specifications required for regulatory use (i.e. related to essential requirements and mandated by the Commission) and other technical specifications has been clarified, and guidelines for their preparation by CEN technical committee have been prepared. And the problem of overlapping directives, involving different approaches to CE marking, is now better understood and arrangements have been developed for its solution.

For the Commission, its Standing Committee on Construction, and the CEN/EOTA technical committees that have to prepare many hundreds of technical specifications, there remain more than a few unresolved issues which are discussed further in Chapter 12.

FURTHER READING

The Construction Products Directive of the European Communities, *Building* Technical File Special, Building (Publishers) Limited, London E14 9RA (June 1989).

Gives full text, reviews long-term implications for the construction professions; the route to European standardization; the principal parties – and some major tasks; contents of the Construction Products Directive, with index; essential requirements; European Standards; European Technical Approvals; Interpretative Documents; the CE Conformity Mark; initial type testing and testing of samples; factory product control. Appendices: useful addresses; technical committees and BSI staff membership; European construction product certification schemes; short glossary of essential terms and names.

G. Blachère, *Building principles*, EEC Directorate-General Internal Market and Industrial Affairs EUR 11320 (1987).

Prepared for the Commission as an introduction to many of the principles on which 89/106/EEC was based.

Building Research Establishment Reports:

BR 177: *The Construction Products Directive of the European Communities. Draft Interpretative Documents: Protection against noise*

BR 179: *The Construction Products Directive of the European Communities. Draft Interpretative Documents: Hygiene, health and the environment*

BR 180: *The Construction Products Directive of the European Communities. Draft Interpretative Documents: Energy economy and heat retention*

BR 181: *The Construction Products Directive of the European Communities. Draft Interpretative Documents: Mechanical resistance and stability*

BR 182: *The Construction Products Directive of the European Communities. Draft Interpretative Documents: Safety in use*

BR 102: *Electromagnetic compatibility requirements for microelectronics in building services: a proposed standard*

(Building Research Station, Watford, 1987)

Since these reports, a number of versions of the Interpretative Documents have been issued by the Commission, the current position being reported in DoE *Euronews Construction*.

Building Research Establishment, BRE Digest 397, *Standardization in support of European legislation: what does it mean for the UK construction industry?* (replaced Digest 376).

Statutory Instruments 1991 No. 1620 Building and Buildings
The Construction Products Regulations 1991 (HMSO, London, 1991).

Made under the European Communities Act, the Regulations implement the Construction Products Directive in the UK. The Regulations are in four Parts: Part I Preliminary, including Interpretation; Part II

Requirements relating to construction products; Part III Enforcement of Part II; Part IV Miscellaneous, including duties of enforcement officers.

From time to time, the European Commission's Standing Committee for Construction has issued guidance papers on the implementation of 89/106/EC. The papers are not to be regarded as legal interpretations of 89/106/EEC, and therefore are not judicially binding. The papers are primarily of interest to those in Member States, and in European bodies like CEN and EOTA, who are involved in giving effect – legally, technically or administratively – to 89/106/EEC.

European standards and technical specifications

9

Technical specifications: totality of the technical prescriptions contained in particular in the tender documents, defining the characteristics required of a work, material, product or supply, which permits a work, material, product or supply to be described in a manner such that it fulfils the use for which it is intended by the contracting authority; these technical prescriptions shall include levels of quality, performance, safety or dimensions, including the requirements applicable to the material, the product, or to the supply as regards quality assurance, terminology, symbols, testing and test methods, packaging, marking or labelling.

They shall also include rules relating to design and costing, the test, inspection and acceptance for works and methods or techniques of construction and all other technical conditions which the contracting authority is in a position to prescribe, under general or specific regulations, in relation to the finished works and to the materials or parts which they involve.

Article 1, EU Directive 83/189/EEC,
as amended by 88/182/EEC

9.1 STANDARDS FOR A SINGLE MARKET

The Single European Market only becomes a reality when manufacturers work to common technical standards against which regulators accept as statements of technical requirements in legislation, and public authority purchasers specify. The New Approach to a Single Market is, therefore, based on the following principles:

- Where European legislation refers to technical specifications, they usually mean harmonized European standards.
- Harmonized European standards in general will be written in performance and not prescriptive terms.
- Standards may specify different levels of performance depending on intended end use.
- Those responsible for national regulations may select levels or classes of performance within a harmonized European standard, related to local conditions and existing regulatory requirements.

Governments and public bodies cannot, without very good reasons, specify for works, or supply of services covered by a 'procurement' directive, a performance related to essential requirements that goes beyond that expressed in an identifiable grade or level in relevant harmonized European standards.

A CE mark, placed on a product by a manufacturer, will demonstrate to officers responsible for enforcement of national regulations that one or more essential requirements are satisfied in a specified way, and that the product can therefore be placed on the market.

The fact that a product carries a CE mark only signifies that, if used correctly, it satisfies relevant regulations. It does nothing more. It is not an indication that the product is of a higher level of performance than that required in regulations. It does not indicate that a product satisfies requirements set out in a non-harmonized standard, or in any technical specification outside the regulatory system.

For these reasons, preparation of New Approach harmonized European standards is supported with Commission funds under mandate to CEN/CENELEC: to underpin technically the New Approach directives; and to be the basis of technical specifications through which all who compete for public contracts do so on level terms.

As they are the principal route to demonstrating conformity with the provisions of European technical legislation, the Single Market cannot be complete until sufficient and adequate European standards are available: an event still some way off. Until then, European legislation aimed at opening up procurement of services, supplies and works throughout the Union cannot be fully effective, neither can there be, for example, CE marking of construction products under 89/106/EEC. It does not mean that this legislation is a dead letter – there are transitional measures available – but it does mean there are gaps that need to be filled rather than patched.

The Single European Act requires European legislation not to lower the protection available in national legislation, usually provided in Member States by reference to national standards. For construction products, protection under 89/106/EEC is given through listing of essential requirements for works transformed through the six Interpretative Documents, which serve to support mandates given to CEN/CENELEC for the preparation of harmonized European standards.

European standards will usually be in two parts:

1. a harmonized part related to one or more essential requirements and therefore to health and safety requirements of national regulations, the costs of preparation being borne by the Commission; and
2. a non-harmonized part covering non-regulatory, usually commercial, requirements whose preparation costs fall on European industry directly or through contributions to national standards bodies.

In preparing a European standard, a CEN technical committee will be helped by the Commission's mandate, which will identify which of the essential requirements are relevant. Against relevant sections of the Interpretative Document or Documents, a committee may abstract and bring together in a single document the best from national standards. To replace national standards is no easy task. It holds out opportunities for European industry; but also poses problems. Like all standards-making, the process must be continuous and dynamic. But to reduce many different national sets of standard requirements to one European set is not easy. When achieved, the result should benefit regulators, consumers and suppliers alike. As EFTA countries are also participating in the work, there is little risk of a weakening of consumer protection or lowering of safety standards.

As many as 2500 European construction standards are likely to be in place by the end of the millennium, largely through implementation of 89/106/EEC: a fact that underlines the urgent need to progress European standards in an orderly manner. To ensure this, a CEN Programming Committee on building and civil engineering has been at work for some time. BTS1, the CEN Sector Technical Board for Building and Civil Engineering, and its TC Technical Committees, are responsible for the work in the construction sector.

Two general issues, important for standards-making as a whole, are the format of a model European product standard, and guidelines for inclusion in product standards of appropriate conformity evaluation procedures, including matters such as factory product control and sampling procedures. Here CEN technical committees, mandated to prepare European standards, have to bear in mind that 89/106/EEC, against which the harmonized parts of European standards are to be drafted, have to meet criteria that sometimes are conflicting; for example:

- the effect of products in use on health and safety of works;
- the effect of variability of a product's characteristics on its serviceability, and on that of the construction in which it is incorporated;
- the susceptibility of a particular product to defective manufacture;
- the requirement that 'the least onerous procedure consistent with safety be chosen'; and
- the needs of public purchasers, who need assurance that levels of technical performance are not eroded.

9.2 GLOBAL FEATURES OF STANDARDS

How best to deal with the horizontal or global features of standards work, such as common procedures for taking and testing samples, and the vertical or sectoral features, such as conditions of storage and use, is another issue. It is convenient to standardize the form in which standards are presented, and for standards-making to follow common rules. There are benefits in using terms that have similar meanings in different standards, and in adopting similar procedures for matters such as attestation of conformity, testing and sampling. Carried to an extreme, however, their adoption – especially where, for example, they have an impact on factory control procedures – can place unnecessary burdens on suppliers. For example, some sectors of industry, for commercial and technical reasons, operate continuous processes of manufacture; other make their products in batches. For this reason, there are limits to the horizontal approach, and there is a need to consider also the needs of different sectors of industry.

Like other national standards bodies, BSI, as a member of CEN/CENELEC, takes an active part in the development of the CEN programme and in the work of CEN technical committees, where UK representatives are selected and briefed by the appropriate BSI technical committees. They are assisted in reaching a consensus on such issues by an advisory Conformity and Quality Panel for Construction Products B/–/2 under B/– (BSI Standards Board for Building and Civil Engineering).

Much of the day-to-day work of standards-making is carried out in English, yet the terms used in European standards and legislation are extending the language. New phrases are entering the language, possibly adapted from French. A new significance, and greater precision, are having to be given to terms and phrases, and new acronyms used as shorthand for the convenience of regular participants. To help other readers, a glossary of English words and phrases used in regulations, standards and codes, quality assurance, testing and certification and definitions, where possible taken from an authoritative source, can be found in Part Two.

9.3 CATEGORIES OF EUROPEAN STANDARDS

As Box 9.1 shows, there are three categories of European standards. For European construction products legislation, harmonized Category B, or harmonized parts of Category B European standards, with associated Category Bh standards, are of first importance. For constructors, and especially for designers, Category A are likely to be of greater interest as they replace design, installation and workmanship standards, or, like the structural Eurocodes, exist at present as ENVs (European prestandards) side by side with national standards.

9.3.1 Harmonized European standards (HENs)

Being linked to the essential requirements, HENs will acquire a quasi-legal status. As such, when available, they will determine the eligibility of a construction product to be CE marked. Where a HEN has been issued, it must be used in preference to other standards as a technical specification for works and supplies covered by European procurement regulations. Although constructors responsible for works not covered by procurement regulations are not obliged to use a HEN as

BOX 9.1 CATEGORIES OF EUROPEAN STANDARDS

CATEGORY A

Design, installation and execution standards (e.g. structural Euro-codes).

Although this category of standards, which equates to codes of practice, plays a key role in the national building legislation of Member States, because the construction product route was adopted to achieve harmonization, preparation of this category of ENs is receiving low priority.

Category B

Product standards of two types:[a]

* HENs (harmonized European standards) mandated by the Commission for those characteristics subject to essential requirements listed in 89/106/EEC Annex I.
* ENs (non-harmonized European standards) for characteristics other than those subject to Annex I essential requirements.

Category Bh

'Horizontal standards' for methods of test and measurement, which span a number of products.

[a] products standards should, in general, be written in performance rather than prescriptive terms

a technical specification, they are likely to do so as European standards replace national standards, particularly in building regulations and similar documents.

With the introduction of harmonized European standards, it will be necessary to distinguish in specifications between properties that relate to essential requirements, and those relating to other 'consumer' requirements. The form in which a category B standard product

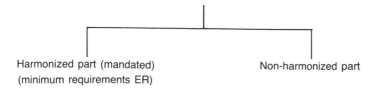

Harmonized part (mandated)   Non-harmonized part
(minimum requirements ER)

Figure 9.1 The two parts of a Category B product standard.

standard is likely to be published will therefore, as already described, consist of a harmonized part directed at the regulatory sector and a non-harmonized part covering matters like colour, appearance and usability (Figure 9.1).

A HEN can be prepared only under 'mandate': that is, a request from the European Commission to the European standards-making bodies CEN/CENELEC. As the work of preparation is financed by the Commission, it will only be authorized if one or more Member States have an Essential Requirement provision in their building regulations. Because preparation of a HEN arises from a requirement in Article 4 of 89/106/EEC, associated with the harmonized standard will be a specification of the method of attestation: that is, how a product must be tested and/or certified.

9.3.2   Non-harmonized European standards (ENs)

ENs cover aspects of performance other than those related to essential requirements. They will serve as technical specifications for contracting authorities specifying requirements for works or supplies not covered by the regulations, taking preference over national and other standards.

Preparation of an EN or a draft European standard (ENv) may be initiated by a CEN committee, a CEN member standards body such as BSI, or a European professional or trade organization. Alternatively, it may be a modification of an ISO standard. As ENs are not mandated, their preparation will have to be funded by CEN and its member bodies or their industry subscribers, and not by the Commission. Drafting of non-harmonized European standards will therefore have a lower priority unless there is a clearly expressed demand from European industry.

The status of European standards is a matter of concern to designers, builders and their clients. A British Standard, although a voluntary document, when used in contracts, is regarded in a UK court of law as representative of the state of the art in an area of construction technology. In other countries, national standards adopted in an official journal have at least the same if not a higher status. European standards, whether or not they result from an EU Mandate, are likely to be of similar importance when they replace national standards. Some Category A European standards will take on a formal status when called up by national building regulations, and will have an important place in public procurement legislation.

As testing, inspection and certification procedures could equally give rise to technical barriers, they also must be harmonized, although the level of conformity attestation required in a standard could differ depending on end use.

9.4 CONFORMITY ATTESTATION PROCEDURES

Regulators and specifiers want assurance that products on the market conform to relevant technical specification or specifications. They can be assured in more than one way. Some ways are more reliable in the information provided; some are less onerous on suppliers.

One way is to let the supplier state in technical literature, in commercial documents or otherwise that a product is 'up to standard'. A regulator or specifier who suspects that it is not up to standard can have it tested: not a very practical procedure where the product has already been built in. If the test shows that the product is not up to standard, there may be a contractual claim or even prosecution under trades description legislation; but if it is as claimed, there is the matter of who pays for the tests.

A more efficient way – less costly for all involved in the long run – is along the quality assurance route. It places responsibility for ensuring that a product is up to standard on the supplier, who assures regulators and specifiers that his system of manufacture is properly managed, with appropriate internal controls, and that the system has been audited and is under the surveillance of a competent, independent auditor. This is the route set out in the BS 5750 quality system series.

Unfortunately, when Commission officials were drafting proposals for 89/106/EEC in the 1980s, possibly from lack of experience they ignored the quality assurance route. Instead, they proposed an approach to conformity attestation largely based on Federal German practice. It has proved difficult to put into operation, mainly because of the difficulty, in a requirement in 89/106/EEC Article 13.4 that the procedure chosen for a harmonized European standards should be 'the least onerous possible procedure consistent with safety', of reconciling the desire of suppliers to avoid costs regarded as onerous and regulators' concern for safety.

There are two different needs in the assessment, certification and approval of building products:

- checking conformity of the attributes of a product with the technical specification given in the relevant product standard or standards, using specified tests and other evaluation procedures (e.g. the Kitemark and Testguard approach); and
- appraisal of suitability for intended use, usually of new products or established products in a new design situation, possibly against agreed guidelines (the *agrément* approach).

At present, variations of these two approaches may be found in most Member States. There are, for example, in France a number of marking schemes indicating conformity to standards issued by AFNOR, the French national standards body, and *Avis technique*, a development from French *agrément*. In Germany, there are extensive arrangements for testing, certification and marking against national DIN standards by official and recognized industrial testing laboratories, and official arrangements for the acceptance of new products and systems – the *zulassung* approach. In the Netherlands, the two approaches may be used together in a *certificaat-met-attest*, which indicates conformity to attributes in a Dutch standard and assessment of performance in use.

9.5 CONFORMITY ATTESTATION UNDER 89/106/EEC

Attestation of conformity to technical specifications is central to the CE marking of construction products. Chapter V 'Attestation of conformity' lays down the procedure and defines responsibilities for attestation of conformity with requirements of a technical specification. Article 13.1 defines the responsibility of a manufacturer as follows:

> The manufacturer, or his agent established in the Union, shall be responsible for the attestation that products are in conformity with the requirements of a technical specification within the meaning of Article 4.

The general principles are set out in Article 13.3:

> The attestation of conformity of a product is dependent on:
>
> (a) the manufacturer having a factory production control system to ensure that product conforms with the technical specifications, or
> (b) for products indicated in the relevant technical specifications, in addition to a factory production control system, an approved certification body being involved in assessment and surveillance of the production control or of the product itself.

Annex III.1 lists the methods required for attestation of conformity:

> (a) initial type-testing by the manufacturer or an approved body;
> (b) testing of samples taken at the factory in accordance with a prescribed plan by the manufacturer or an approved body;
> (c) audit-testing of samples taken at the factory, on the open market or on a construction site by the manufacturer or an approved body;
> (d) testing of samples from a batch which is ready for delivery, or has been delivered by the manufacturer or an approved body;
> (e) factory production control;
> (f) initial inspection of factory and of factory production control by an approved body;
> (g) continuous surveillance, judgement and assessment of factory production control by an approved body.

The preferred systems of conformity attestation are summarized Box 9.2; it is for the Commission, after consultation with the Standing

BOX 9.2 SYSTEMS OF ATTESTATION OF CONFORMITY TO
TECHNICAL SPECIFICATIONS

Certification by an approved body

Tasks for the manufacturer:
* operate required factory production control
* carry out any further testing of samples taken at the factory in accordance with a prescribed test plan

Tasks for the approved body
* carry out initial type-testing of the product
* carry out initial inspection of factory and of factory production control
* maintain continuous surveillance, assessment and approval of factory production control
* possibly, undertake audit-testing of samples taken at factory, on the open market or on a construction site

Manufacturer's declaration

First possibility, involving an approved body
Tasks for the manufacturer:
* carry out initial type-testing of the product
* operate required factory production control
* carry out any further testing of samples taken at the factory in accordance with a prescribed test plan

Tasks for the approved body:
* certification of factory production control on the basis of initial inspection of factory and of factory production control
* possibly, maintain continuous surveillance, assessment and approval of factory production control

Second possibility, involving an approved laboratory
Tasks for the manufacturer:
* operate required factory production control

Tasks for the approved laboratory:
* carry out initial type testing of the product

Third possibility
Tasks for the manufacturer:
* carry out initial typetesting
* operate required factory production control

Committee, to decide which of these procedures should be followed by a manufacturer, and to indicate the procedure chosen in its mandate to CEN/CENELEC, who will record the decision as a requirement of the relevant harmonized European standard.

In making its choice, the Commission has to take into account:

- the importance of the part played by the product with respect to the essential requirements, in particular to those relating to health and safety;
- the nature of the product;
- the effect of the variability of the product's characteristics on its serviceability; and
- the susceptibility to defects in the product manufacture, in accordance with the particulars set out in Annex III.

It is also charged with giving preference in each case to 'the least onerous possible procedure consistent with safety'.

Here, the Commission has to balance freedom of manufacturers to develop and innovate, and the need to promote competition between materials, systems, and sectors of manufacturing industries, and permit sufficient flexibility to facilitate improvements in production techniques, against the need to protect users, particularly where use may involve risks to the health and safety of occupants and users of buildings, or, in the case of civil engineering works, a disastrous structural failure, from products whose performance in use is less easy to predict, and from producers who are inexperienced in the sector of industry being considered, or are prone to take short cuts in order to maintain profitability. The task is not easy, and cannot depend on manufacturing interests alone. The Commission should therefore draw on international work that has gone on, and is currently in progress within bodies like CIB and RILEM, and on experience of members of ENBRI (the European Network of Building Research Institutes).

Article 13.5 deals with the special case of one-off products:

> In the case of individual (and non-series) production, a declaration of conformity ... shall suffice, unless otherwise provided by the technical specifications, for products which have particularly important implications for health and safety.

In the case of a manufacturer's Declaration of Conformity, variations of the manufacturer's declaration may involve an initial type test by an approved laboratory, or certification of the factory production control system by a third party. The Commission will decide which variant, if any, is applicable to a particular product group. The UK Construction Products Regulation 5 requires the manufacturer to keep the Declaration of Conformity available for inspection by the enforcement authorities for a period of at least 10 years. If a product needs to be independently tested and/or certified, the decision taken by the Commission will be published in the *Official Journal of the EU*

(C series), and it will be up to individual trade associations to keep track of such decisions and inform their members. In the UK, the BRE will hold information and the decisions are expected to be publicized through the trade press, and in *Euronews Construction*.

9.6 APPROVED BODIES

The systems of attestation of conformity summarized in Box 9.2 assign tasks to three kinds of approved bodies: certification bodies, inspection bodies and testing laboratories, all to be designated by their Member State. 89/106/EEC Annex III.3 defines their respective functions:

(a) certification body, defined as
an impartial body, governmental or non-governmental, possessing the necessary competence and responsibility to carry out conformity certification according to given rules of procedure and management;

(b) inspection body, defined as
an impartial body, having the organization, staffing, competence and integrity to perform according to specified criteria functions such as assessing, recommending for acceptance and subsequent audit of manufacturers' quality control operations. and the selection and evaluation of products on site or in factories or elsewhere, according to specific criteria;

and

(c) testing laboratory, defined as
a laboratory which measures, examines, tests, calibrates or otherwise determines the characteristics or performance of materials or products.

In distinguishing between the three types of body, the Annex notes that for certification, inspection and testing may be undertaken by the same body, but where inspection and testing are undertaken by separate bodies they do so on behalf of a certification body.

Member States have to task of designating these bodies, which must fulfil the following minimum conditions:

1. availability of personnel and of the necessary means and equipment;
2. technical competence and professional integrity of personnel;
3. impartiality, in carrying out the tasks, preparing the reports, issuing the certificates and performing the surveillance provided for in the Directive, of staff and technical personnel in relation to all circles, groups and persons directly or indirectly concerned with construction products;
4. maintenance of professional secrecy by personnel;

5. subscription of a civil liability insurance unless that liability is covered by the State under national law.

Fulfilment of the first two conditions has to be verified by the competent authorities of Member States, who are required to notify the Commission and other Member States of the names and addresses of the bodies designated. The Commission has the task of ensuring that a list, and any subsequent amendments, are published in the C series of the *Official Journal of the European Union*. Further detailed guidance on the designation of approved bodies is given in EU SCC Guidance Paper 6.

For the UK, the Department of the Environment invited organizations interested in being designated as approved bodies to register their interests in November 1990, entrusting the task of registration to BRE. To advise the Department on operational details of the designation process, a Designation Steering Group has been set up and Guidance Notes issued by BRE European Unit.

9.7 NATIONAL STANDARDS DURING THE TRANSITIONAL PERIOD

Until a HEN, a European technical approval or an EU-recognized national standard is available, suppliers cannot legally CE mark their products. During a transitional period, they will continue to have to ensure that their products conform to their national standards, or the national standards and other technical specifications of the Member State to which they plan to export.

Usually this means no change from the current situation: conformity to the relevant standard or standards of the importing country, the status of which varies greatly. In Federal Germany, for example, DINs are the basis of a comprehensive regulatory system, while in countries like Denmark and the Netherlands their 'deemed to satisfy' status is not that different from the UK, where some but by no means all BSs are referred to in relevant Approved Documents. In France, and countries like Spain with similar administrative arrangements, official adoption (homologation) and withdrawal (annulation) of a national standard is noticed in that country's official journal.

Under European procurement legislation, the situation is somewhat different as, in the absence of a European standard, contracting authorities are able to refer in contract documents to national standards provided they make clear that any other technical specification is acceptable – if equivalent.

9.8 STRUCTURAL EUROCODES: A PARTICULAR KIND OF EUROPEAN STANDARD

The structural Eurocodes are examples of ENVs (European pre-standards) which, after experience in use, will be adopted by CEN as Category A European standards. Their initial preparation was on the direct initiative of the Commission, which sponsored preparatory work begun before the New Approach to a Single Market, one aim being to provide technical reference documents for use in the preparation of products standards under 89/106/EEC.

Other uses were also intended:

- as documents that could be used to show compliance with national building regulations;
- as documents that could be used to justify a design option in connection with tendering for work under EU procurement rules; and
- as tools for strengthening the position of European constructors in the world markets.

The ideal of a set of European structural codes had merits. To have available a coherent and meaningful set of European structural codes was regarded as a first step to a more uniform system of regulation in the interests of safety in buildings and the creation of a Single Market for construction. The process of developing that system has been slow: slower than many thought when the journey was embarked on in the mid-1980s. Originally to be issued by the Commission, they have been drafted by panels of independent experts appointed directly.

Following a Commission policy change in the late 1980s, responsibility for Eurocodes was transferred to a CEN technical committee, CEN/TC 250. From 1990 onwards CEN began to make Eurocodes available to CEN member countries as ENV documents, it being anticipated that each member country would also issue a National Application Document (NAD), giving guidance on national application with definitive figures for national regulated safety values and a schedule of supporting national references standards. The state of preparation and publication is set out in Box 9.3. Progress with development and approval, and dates and arrangements for publication, are reported in *Euronews Construction*.

In the UK, the Eurocode prestandards are being published by BSI as DDs (Drafts for Development) incorporating the relevant NAD giving essential guidance on use in the UK. As other member countries have different names and procedures for publishing drafts, publication of Eurocodes in each country's national system will not necessarily be simultaneous, and there may be variations in references in NADs. Again, the status of Eurocodes in national building regulations may differ as not all countries have adopted the UK Approved Document system, which allows a central regulating authority, like DoE, to identify available ENVs, when used with their NADs, as appropriate guidance for the design of buildings. To help designers,

BOX 9.3 THE STRUCTURAL EUROCODES: STATUS REPORT
          (AUGUST 1993)

ENV 1991	Basis of design (approved for publication, May 1993)
ENV 1991-2-1	Densities, Self-weight and imposed Loads
ENV 1991-2-2	Snow loads
ENV 1991-2-7	Actions from structures exposed to fire
ENV 1991-4	Actions in silos and tanks (accepted by CEN/TC250/SC1 for publication, June 1993)
ENV 1991-2-3	Wind actions (final approval deferred)
ENV 1991-3	Traffic loads on bridges
ENV 1991-3-4	Rail traffic actions (both likely to be accepted in December 1993)
ENV 1992	Design of concrete structures (as Eurocode 2: issued by BSI as Draft for Development with National Application Document)
ENV 1992-1-3	Precast concrete elements and structures
ENV 1992-1-4	Lightweight aggregate concrete
ENV 1992-1-5	Unbonded and external pre-stressing tendons
ENV 1992-1-6	Plain concrete structures (likely to be published early 1994)
ENV 1992-1-2	Structural fire design (still being considered)
ENV 1993	Design of steel structures: General rules and rules for buildings (likely to be published early 1994: as Eurocode 3 issued by BSI as Draft for Development with NAD)
ENV 1993-1-3	Cold-formed thin-gauge members and sheeting, with two Annexes (likely to be published early 1994)
ENV 1993-1-2	Structural fire design (still being considered)
ENV 1994	Concrete and steel composite construction (Eurocode 4) Part 1: General rules and rules for buildings (approved)
ENV 1994-1-2	Structural fire design (to be approved in 1993)
ENV 1995	Timber (Eurocode 5)
ENV 1995-1-1	General rules and rules for buildings (likely to be published early 1994)
ENV 1996	Masonry (Eurocode 6)
ENV 1996-1-1	General rules and rules for buildings
ENV 1996-1-2	Structural fire design (both still to be accepted)
ENV 1997	Geotechnical design (Eurocode 7)
ENV 1997-1-1	General rules (accepted in June 1993)
ENV 1998	Earth-resistant design of structures (Eurocode 8)
ENV 1998-1-1/3	General requirements, rules for buildings and specific rules for various materials (under development)
ENV 1999	Aluminium (work on development approved by CEN)

BSI is publishing national applications of Eurocodes in France and Germany as well as BS DD ENVs. An account of the scope and purpose of the structural Eurocodes was the subject of a special supplement to the July 1992 *Euronews Construction*.

During 1994, the Committee for the Structural Eurocodes CEDN/TC250 set 1 January 2000 as the key target date by which all the EN Eurocodes relating to the design of structures for buildings shall have been published. It also decided that the following parts of Eurocodes 1, 2, 3 and 4 will be published before the end of 1998:

- Eurocode 1: EN 1991-1, Basis of design; EN 1991-2-1, Actions: density, self-weight, and imposed loads; EN 1991-2-3, Actions: snow loads; EN 1991-2-4, Actions: windloads.
- Eurocode 2: EN 1992-1-1, Design of concrete structures: general rules and rules for buildings.
- Eurocode 3: EN 1993-1-1, Design of steel structures: general rules and rules for buildings.
- Eurocode 4: EN 1994-1-1, Design of composite concrete and steel structures: general rules and rules for buildings.

Although Eurocodes relate to the first of the six essential requirements, *Mechanical resistance and stability*, they do not fit neatly into a New Approach directive like 89/106/EEC, and they cannot be said to facilitate directly removal of barriers to trade within the Union. Rather, they provide an alternative to national documents for (at present) voluntary use by national authorities. For example, the 1992 DoE Approved Document A identifies the two DDs as appropriate design guidance; and it is anticipated that they will still serve to 'justify an alternative design in connection with a tender invited by a public works authority under the Public Works Directive'.

In the longer run, as resources become available, similar Eurocodes will be needed in such areas as fire safety, protection against noise, and energy conservation systems. When this happens, the considerable experience gained in the drafting and assessment of the structural Eurocodes will be invaluable.

FURTHER READING

Supplemented by information in DoE's *Euronews Construction* and in the monthly *BSI Update*, constructors, manufacturers and regulators learn about work on European standards through representative membership of a BSI technical committee, or through their professional or trade association. Information is of three kinds: work programmes for drafting ENs; progress achieved in drafting ENs; replacement of BSs by ENs. Manufacturers have an ongoing interest; but only when familiar and less familiar national standards are phased out and European standards take their place will the interest of constructors and regulators be aroused. This has already started in some fields and will become of increasing importance during the 1990s and beyond as European standards are transformed into national standards – normally within 6 months – regardless of the way in which their national standards body voted, and any national standard that conflicts with the European standards. *BSI Update* reports new international and European, as well as British, standards work. The lists are comprehensive, covering all technologies, and only part is of direct relevance to construction. A second source is OJ, the *Official Journal of the European Union*, where harmonized European standards have to be reported before being transposed as national standards– in the UK as BS EN.

The European Commission has issued, as CONSTRUCT 94/118, a Communication dated 6 October 1994: *Standardization and Implementation of the Construction Products Directive*. It gives a general view of the implementation by the Commission for the finalization of the 33 mandates for harmonized construction product standards; attempts to distinguish between descriptive and performance standards, and between harmonized and voluntary standards; and discusses 'fitness for intended use'.

There is an explanation of the development and status of the structural Eurocodes in a *Euronews Construction* special supplement dated 2 July 1992. There is further information on the structural Eurocodes in later issues.

The Procurement and other Directives 10

Directive: a set of rules promoted by the Commission, and adopted by the Council of Ministers after appropriate consultation, binding in respects of results to be achieved; but left to individual Member States to implement by an appropriate administrative, or legislative procedure.

DTI, *The Single Market: the facts* (1992)

89/106/EEC is not the only item of European legislation affecting construction, although it may be the item given most attention at present because, until it is fully operational, national standards for products cannot be replaced by harmonized European standards, and the special status it gives to the *agrément* concept under the guise of European technical approvals cannot be realized.

Directives likely to affect construction are listed in Table 10.1. Up-to-date lists, with information on progress of implementation, are published in *Euronews Construction*. The Department of Trade and Industry gives additional information on EU directives in its 'Single Market' series, which also give the central government contact point through which further information may be obtained.

Most of these directives follow the pattern of New Approach directives and refer to European technical specifications where there is a need to demonstrate that essential requirements are satisfied, although the way reference to the technical specification is made, and the conformity procedures required, vary.

Table 10.1 groups the directives under a number of heads:

- public procurement;
- mutual recognition;
- other directives;
- health and safety;
- Construction Products Directive;
- marketing and use;
- consumer affairs.

All are directed at one of two objectives: promoting freedom of establishment and trade within the EU; and safeguarding the personal health and safety of EU citizens.

Table 10.1 European legislation affecting construction (at November 1994)

Title	European Commission reference	State of play	Official Journal reference
Public procurement			
Works Directive (consolidated)	932/37/EEC	Implemented – 21.12.91	L199 – 09.08.93
Supplies Directive	77/62/EEC	Implemented	L13 – 15.01.77
Supplies Directive	80/767/EEC	Implemented	L215 – 18.08.80
Supplies Directive	88/295/EEC	Implemented – 21.12.91	L127 – 20.05.88
Supplies Directive (consolidated)	93/36/EEC	Adopted – 14.06.93	L199 – 09.08.93
Compliance Directive	89/665/EEC	Implemented – 21.12.91	L395 – 30.12.89
Utilities Directive	90/531/EEC	Implemented – 13.01.93	L297 – 29.10.90
Utilities Directive (Consolidation/services)	93/38/EEC	Adopted – 14.06.93	L199 – 09.08.93
Utilities Remedies Directive	92/13/EEC	Implemented – 13.01.93	L76 – 23.03.92
Services Directive	92/50/EEC	Implemented – 13.01.94	L209 – 24.07.92
Mutual recognition			
Architects Directive	85/384/EEC	Implemented – 21.10.87	L223 – 21.08.85
1st General Directive	89/48/EEC	Implemented – 17.04.91	L19 – 24.01.89
2nd General Directive	92/51/EEC	Adopted – 18.06.92	L209 – 24.07.92
Other directives			
Hot Water Boilers	92/42/EEC	Implemented 01.01..94	L167 – 22.06.92
Proposed Packaging Directive	COM(93)416	Under discussion	Document available from European Commission
Electromagnetic Compatibility Directive	89/336/EEC	Implemented – 03.05.89	L139 – 23.05.89
ECD Amendment	92/31/EEC	Implemented – 28.05.92	L126 – 12.05.92
General Product Safety Directive	92/59/EEC	Adopted – 29.06.992	L228 – 11.08.92
Machinery Directive	89/392/EEC	Implemented – 30.11.92	L183 – 29.06.89
Machinery Directive 1st Amendment	91/368/EEC	Implemented – 30.11.92	L198 – 22.07.91
Machinery Directive 2nd Amendment	93/44/EEC	Implemented – 14.06.93	L175 – 19.7.93

Table 10.1 *Continued*

Title	European Commission reference	State of play	Official Journal reference
Lifts Directive	COM(92)35	Under discussion	C62 – 11.3.92
Pressure Equipment Directive	COM(93)319	Under discussion	C246 – 9.9.93
Health and safety			
Construction Sites	92/57/EEC	Adopted – 24.06.92	L245 – 26.08.92
Framework Directive	89/391/EEC	Implemented – 01.01.93	L183 – 29.06.89
Workplace Directive	89/654/EEC	Implemented – 01.01.93	L393 – 30.12.89
Display Screen Equipment	90/270/EEC	Implemented – 01.01.93	L156 – 21.06.90
Manual Handling of Loads	90/269/EEC	Implemented 01.01.93	L156 – 21.06.90
Use of Work Equipment	89/655/EEC	Implemented – 01.01.93	L139 – 30.12.89
Temporary Workers Directive	91/383/EEC	Implemented – 01.01.93	L206 – 29.07.91
Safety Signs Directive	92/58/EEC	Adopted – 24.06.92	L245 – 26.08.92
Use of Personal Protective Equipment	89/656/EEC	Implemented – 01.01.93	L393 – 30.12.89
Extractive Industries:			
Mines and Quarries Directive	92/104/EEC	Adopted – 03.11.92	L404 – 31.12.92
Explosives Directive	93/15/EEC	Adopted – 05.04.93	L121 – 15.05.93
Physical Agents	COM(92)560	Under discussion	C77 – 18.03.93
Construction Products Directive	89/106/EEC	Implemented – 27.12.91	L40 – 11.02.89
Marketing and use			
Asbestos	91/659/EEC	Implemented – 01.01.93	L363 – 21.12.91
Pentachlorophenol	91/173/EEC	Implemented – 31.01.93	L85 – 05.04.91
Consumer affairs			
Unfair Terms in Consumer Contracts Directive	93/13/EEC	Adopted – 05.04.93	L95 – 21.04.93

10.1 EUROPEAN PUBLIC PROCUREMENT LEGISLATION

Central to the concept of the Single European Market is the opening of all but small public contracts to competition. Together, major public contracts amount to about 15% of Europe's GDP, but often they are closed to open award. European procurement legislation dates from the 1970s, but its present state of effective completion started in the late 1990s. With the Utilities Services Directive coming into force in July 1994, an important stage in the Single Market programme will have been reached.

The way in which public sector tenders for building and civil engineering work and for supply of goods and services are sought, the tender documents and technical specifications used, and the manner in which contracts are awarded have, up to now, varied from one Member State to another. The forms and procedures used have, in part, been determined by the national legal system – Roman or Anglo-Saxon – on which contract law is based, in part by the administrative machinery of central governments, and in part by the degree of financial independence of local authorities and state enterprises. The framework within which procurement takes place is also influenced by the nature of national and local construction industries, including the role played by professional architects, surveyors and civil engineers as agents of the building owner, the financial and management sophistication of contractors, the strengths of general, trade and specialist contractors, and the degree to which work is subcontracted.

In most continental countries, public contracts are structured around a framework of Roman law, but only in France are procedures closely defined in administrative law under the *Code des marchés publics* supported by guidelines issued by a central commission on public contracts. In Germany, contract procedures are specified in VOB (*Verdingungsordnung für Bauleistungen*), which set out standard general conditions for three types of invitations to tender – open, selective and direct award – to which Federal and Länder authorities are required to conform.

Table 10.2 summarizes the various procedures for the award of public contracts commonly met with in some Member States. National procurement practices in Federal Germany, France, Spain, Portugal and Italy are fully described in CIRIA Special Publications, and in Spon's *European Construction Costs Handbook*.

The basic structure of the new European procurement legislation is set out in Table 10.3. It remains influenced to a large extent by the form taken by existing public contracts procedures – open, restricted and negotiated – but within that structure a number of general principles are introduced to promote open competition. They are:

- no discrimination on nationality grounds;
- free movement of goods;

Table 10.2 Procedures for award of public contracts

Award procedure	Description of procedure
Open procedure: version 1 (predominant in Portugal and Greece; frequent in Spain)	1. All interested in a contract as advertised can submit a tender
Open procedure: version 2 (predominant in Italy; frequent in France)	2. All interested in a contract as advertised can apply to be selected, but only those selected can submit a tender
Restricted procedure (usual in Germany & Denmark; predominant in UK)	Only those qualified by public contracting authority may participate
Negotiated procedure (*Verhandlungsverfahren*) (frequent in Belgium and Netherlands; subject to precise regulation in France)	Public authority negotiates with one or more suppliers of its choice

- no quantitative restrictions on imports and exports within the Union, and prohibition of measures having equivalent effect;
- freedom of establishment;
- freedom to provide services;
- transparency in technical specifications;
- for open tenders, clearly expressed conditions, and adequate allowance of time for submissions;
- where tendering is restricted by invitation, clearly expressed qualification criteria, adequate allowance of time for application to be included on the qualification list, and adequate allowance of time for submission of tenders;
- limited use of negotiated contracts, and where used only in specific situations and against clearly expressed conditions (see footnote to Table 10.3);
- provision of awarding contracts in an emergency, and contracts covering periodical works like building maintenance;
- transparency of criteria used in making awards; and
- specification of remedies open to businesses harmed by a breach in procurement rules.

Of particular interest to those drawing up tender and contract documents coming within the scope of procurement legislation is the requirement that technical specifications, defining characteristics of works, products, equipment, supplies or services, must be included in documents, and should permit the result to be fit for the intended use. Where they exist, documents should refer to national standards implementing European standards.

Table 10.3 Tender and award procedures specified in procurement legislation

Contract tender and award procedure	Description of procedure
Open procedure: all interested can submit tender	Contract noticed in OJ
Restricted procedure: all interested can apply to be considered but only a number from list of qualified are invited to submit tender	
Negotiated procedure[a]	Public authority negotiates with one or more suppliers of its choice

[a] Only for use in certain defined cases, and when used reasons must be recorded:
1. In general:
   (a) no tenders received under open or restricted procedure; or
   (b) goods can be supplied only by particular supplier on artistic or technical grounds, or because supplier has exclusive rights
   (c) extreme urgency due to an unforeseen event.
2. For supplies contracts:
   (a) goods only for research and development; or
   (b) additional deliveries sought from a particular supplier for reasons of compatability or on technical grounds.
3. For works contracts:
   (a) additional works cannot be separated from main contract, or are strictly necessary to its later completion; or, under certain specific circumstances, works are a repeat of previous contract.
4. For works contracts with a call for competition:
   (a) where tenders received under open or restricted procedure were irregular or unacceptable; or
   (b) nature of work does not permit overall pricing.

Beyond the requirement to refer to European technical specifications, the three groups of directives – public works, supplies and utilities – have somewhat different specification requirements, as shown in Table 10.4.

It is acceptable, and may be necessary, to specify a requirement additional to those in the relevant European standard, provided it does not change essential requirements in that standard, taking into account any class or level of performance specified, and the requirement is objectively justifiable and proportionate. However, at least for construction products, a higher attestation level, for example third-party certification, than that specified in the standard cannot be required, as this might be ruled as a measure having an equivalent effect as a quantitative restriction on imports.

Where there is no relevant European standard, specifications may be based on the principles of equivalence and mutual recognition of either UK or other published standards. In their absence, contracting authorities must be willing to consider items that meet the required level of performance and are fit for the intended use or purpose.

Table 10.4 Specification requirements

Public Procurement Directive	Specification requirement
Public Works Directive	Comprehensive reference to technical specifications: (a) national specifications implementing European standards[a]; (b) European technical approvals; or (c) common technical specifications[b]
Public Supplies Directive	As 89/106/EEC; but does not refer to European technical approvals
Utilities Directive	More flexible, requiring reference to European specifications where they exist

[a] First national standards based on HENs (harmonized European standards), and then on other European standards.
[b] Defined as technical specifications, laid down in accordance with a procedure recognized by Member States to ensure uniform application in all Member States, which have been published in the *Official Journal of the European Communities*.

In addition to requiring that products meet a technical specification, a specifier may be able to require quality management certification under BS 5750/EN 29000 by a named certification body as long as the term 'or equivalent' is included to allow another 'equivalent' quality assurance route. It is also possible to claim a derogation where compatibility with existing equipment is essential; or costs resulting from departing from previously used technical specifications are unacceptable.

Tender and contract documents must also make clear the criteria used in making an award. They are of two kinds:

1. *Qualification criteria*: The contracting authority under Supplies And Works Directives may require evidence of economic, financial and technical standing of supplier, but only of a kind laid down in the directives; the Utilities Directive may require the supplier to be on an official prequalification list, which must be open.
2. *Criteria to be considered when awarding a contract*: Whether the contracting authority will make the award on price alone; or on a combination of factors such as price, quality and delivery, enabling the purchaser to assess value for money; additionally, for supply of services, whether account may be taken of technical merit, aesthetic and functional characteristics, and technical assistance and services.

Because proposals for procurement legislation were made by the Commission and debated and adopted by the Council of Ministers at

---

BOX 10.1 DEFINITION OF CONTRACTING AUTHORITY

Under Works and Supplies Directives

1. United Kingdom[a]

   (a) Government Departments and other Crown bodies
   (b) Local authorities
   (c) Certain other bodies 'governed by public law' such as education authorities, fire and police authorities, and the national health service authorities; also
   (d) Certain contracts where a contracting authority puts up more than 50% of funding or where the contracting authority seeks Community financial assistance

2. Other Member States

   Management of public business differs traditionally and, within the criteria set out, the list of contracting varies from Member State to Member State and may or may not include certain university and research organizations.

Under Utilities Directive

As well as public undertakings, certain private sector holding 'special or exclusive rights' such as the right to supply or manage a public telecommunication, transport, drinking water, or energy network

---

[a] Bodies covered by these criteria are listed in Annex I of 89/440/EEC. The list is not exhaustive and may be by the Commission after notification by a Member State

---

different times, specific requirements in procurement directives may differ. There are, for example, differences in definitions of contracting authorities, the most important being that given in 90/531/EEC, the Utilities Directive covering the so-called 'excluded sectors' – energy, water, transport and telecoms – where, in addition to central, regional and local authorities and similar bodies in the public sector, utilities in the energy, drinking water, transport and telecoms private sector are defined as contracting authorities, although under a number of modified requirement conditions. One exclusion, affecting defence and police services, is that directed at supplies declared secret or subject to national security measures. Again, procurement legislation may extend to contracting authorities other than central and local government. Coverage can be somewhat different in different Member States, as shown in Box 10.1.

Table 10.5 Definition of thresholds

Contracting authority	Type of contract	Threshold[a]
Public bodies and certain other bodies	Supplies	200 000[b]
	Works	5 000 000
	Services	200 000
Utilities	Supplies	400 000
	Works	5 000 000[c]
	Services	400 000

[a] Values are express in ECU; in May 1993 £1 = 1.26 ECU
[b] For GATT bodies 125 575 ECU
[c] For telecoms 200 000 ECU

Table 10.6 Time limits for tender submissions

Type of contract procedure	Time that must elapse after dispatch of notice[a]
Open	At least 52 days before deadline for tender submission, or 36 days where sufficiently specific periodic notices has been published[b]
Restricted or negotiated	At least 37 days before deadline for receipt of application; and at least 40 days between invitation to tender and deadline for tender submission

[a] Time limit may be extended where it is essential that tenderers visit a site or inspect documents on the spot; under Utilities Directive, parties may mutually agree time limits. Works concessionaires must also advertise contracts above the threshold but need not comply with full requirements of 89/440/EEC.
[b] purchasing bodies under 89/440/EEC must publish a PIN (periodic indicative notice) as soon as possible after a decision to plan work in stages; and under 90/531/EEC a PIN indicating any intended procurement programme for supplies and works over the next 12 months

As well as defining the types of contracting authority covered by the procurement legislation, thresholds are defined as in Table 10.5.

Time limits for tender submissions in open contracts, and for applications to participate and for tender submissions in restricted contracts, are also defined (Table 10.6).

Two types of services are defined in 92/50/EEC, the Services Directive (Table 10.7).

The Treasury is the central government department in the UK for overseeing European procurement legislation. It has issued Consolidated Guidelines on Public Purchasing Policy, and has drafted regulations for implementing the Services Directive. A Single Market Compliance Unit has been established at the Department of Trade and Industry to advise and, where possible, assist UK businesses experiencing barriers to trade that appear to be due to breaches in

Table 10.7 Types of services defined in 92/50/EEC

Priority services: Annex 1A	Residual services: Annex 1B
Maintenance and repair of equipment	Restaurant and hotel
Land transport and courier, air transport	Transport by rail and water
Transport of mail by air, and by land, except by rail	Supporting and auxiliary transport
Financial, accounting, auditing, and bookkeeping	Legal
Management consultancy and related	Personnel
Market research and public opinion polling	Investigation and security
Publishing, printing and advertising	Recreational, cultural and sporting
Sewage and refuse disposal	Health and social
Architectural, engineering, planning and related scientific and technical	Education and vocational educational
Usual telecommunication, computer and related	Services not covered
Building, cleaning and property management	Broadcasting
R&D benefiting exclusively contracting authority for use in own affairs	Certain real estate transactions
	Other R&D

European legislation by other Member States. Information on contracts open, as well as notices appearing in the *Official Journal*, is available from TED (Tenders Electronic Daily) online data service, available through EU's ECHO service, direct or through local EIC (Euro-information Centres).

Not covered by European procurement legislation, public contracting authorities may be subject to rules in the GATT Agreement on Government Purchasing (GAP) made in the context of GATT (General Agreement on Tariffs and Trade). Under these rules, authorities specified in the Agreement are subject to a lower threshold than other public bodies. There is also an obligation to debrief unsuccessful tenderers and tell them why they failed to secure a contract.

10.2 ENFORCEMENT IN THE UK

Enforcement is an important aspect of procurement legislation. There can be deliberate or unwitting breach of rules by a contracting authority, such as:

- breaking up work or supplies into packages of contracts below threshold levels;
- failure to advertise a contract;
- failure to provide full information to firms tendering or considering tendering;
- failure to adhere to time limits set in common rules;
- using discriminatory terms in contract documents, or unfair criteria for qualification of applicants to tender which cannot be met by all qualified tenderers; and/or
- adopting criteria in making an award, or in negotiations afterwards, not specified in the notice calling for tenders or in tender documents.

To obtain redress in the event of a breach of the common rules, procedures are set out in 89/665/EEC, the Compliance Directive, and in 92/13/EEC, the Utilities Remedies Directive. Both prescribe the nature of the minimum remedies to be made available, but leave it to individual Member States to select the legal form and structure of review system best suited to national traditions.

In the UK the legal form adopted is through regulations made under two Statutory Instruments (SI). The regulations provide as follows:

- Contractors and suppliers have recourse to an appropriate court of law – in England and Wales, the High Court; in Scotland, the Court of Session – which has power to grant interim and final injunctions, and award damages, including injunctions and damages against the Crown.
- Aggrieved contractors must lodge a prior complaint with the contracting authority and must make a claim to the court within three months from the time of the alleged breach.
- Only contractors and suppliers harmed by a breach of the rules, or at risk of harm, may seek a remedy.
- Remedies may be in the form of suspension of contract award procedures, setting aside decisions if the contract is not yet entered into, and/or award of damages before or afterwards.

Additional to a degree of uncertainty over the legal definition of 'contracting authority', which is specified differently in the three groups of procurement directives, there are other matters that could obstruct the opening-up of a common European market, such as:

- definition of a security measure;
- definition of urgency by a contracting authority in adopting the negotiated contract procedure;

- the difficulty, particularly for smaller contracting authorities, of knowing what technical specifications are permissible;
- the difficulty, particularly for smaller contractors, of keeping informed about open offers or requests for applications for qualification, in meeting qualification requirements, in studying tender conditions and submitting tender documents in a foreign language; and
- the limited scope of Commission intervention, with the legal form and structure of review systems being left to individual Member States to select.

10.3 OTHER DIRECTIVES: THE PROBLEM OF OVERLAPPING REQUIREMENTS

Directives listed in Box 10.1 may affect construction quality by determining:

- safety of products;
- qualifications, and conditions of employment and establishment of construction professionals;
- health and safety on construction sites, and in other workplaces;
- use of hazardous substances and processes; and
- provisions for environmental protection.

The directives on mutual recognition of qualifications are an important group. The first, the Architects Directive, dates from 1985, the two General Directives being more recent. The Health and Safety Directives, based on a Framework Directive, and, in the UK, implemented under Regulations made by the Health and Safety Commission and enforced by its Executive, make a second group. There are also a number of general safety and environmental protection directives, such as the General Products Safety Directive, the Marketing and Use of Dangerous Substances Directive, and the Urban Waste Water Treatment Directive.

Some of these directives require CE marking to indicate conformity with the requirements of European technical specifications. Proposed by different Directorates in the Commission, not all require identical levels of conformity attestation. This lack of consistency hinders enforcement.

The problem will eventually be overcome when all relevant directives, including 89/106/EEC, are amended to follow guiding principles for conformity assessment based on consistent use of European standards in the EN 29000 and EN 45000 series, and on the promotion of mutual recognition agreements on certification and testing between bodies operating in the non-regulatory sphere.

FURTHER READING

Public Procurement Directives

The Commission has published, in the *Official Journal of the European Union* No C358, an 80-page provisional guide to the EU directives relating to public contracts. Its aim is to help industry understand the reasons for these directives. The guide is in four parts: principles behind public contracts; EU legislation relating to the two directives: public contracts for works, and public contracts for supply of services; method of recourse in cases of non-compliance with the directives; and, financial arrangements for public contracts. In the UK, there is guidance in the DTI *The Single Market* series: *Guide to public purchasing*, 4th edition (March 1992).

The Directives have been implemented by the UK government under:

- The Public Supply Contracts Regulations 1991 (SI 1991 No 2679); and
- The Public Works Contracts Regulations 1991 (SI 1991 No 2680)

    both published by HMSO.

Other directives

A number of the DTI Single Market booklets, mentioned earlier, are devoted to other directives:

Electromagnetic Compatibility – the Directive in brief (May 1992; takes account of amending Directive 92/31/EEC)
Gas Appliances – the Directive in brief (2nd edn, June 1992)
Machinery (October 1991)
Personal protective equipment

Available from DTI 1992, PO Box 1992, Cirencester GL7 1RN, or by telephone from 0181 200 1992.

Euronews Construction of *DoE Construction Monitor* reports progress in the development, adoption and implementation of directives affecting construction.

Achievement of construction quality in six European countries

11

The achievement of quality in construction is a result of many initiatives:

- the aspirations of building owners, their approach to the choice of professional advisers, their ability to identify and express their requirements, and to communicate them to architects and other professional advisers;
- the experience, enthusiasm, professional and technical competence, and management skills of all involved in the design–construction process;
- the ability of regulators to balance architectural and technical innovation against the need to ensure that works do not cause environmental damage, or involve unacceptable risks to safety and health;
- the quality of information for briefing, design work, construction and maintenance, and its effective communication; and
- the way in which research-derived information is communicated to and understood by standards-makers and practitioners, and how the experiences of practitioners are communicated to standards-makers and researchers.

The influence that these matters have will change with changing requirements, attitudes and experiences of individuals and national organizations. All contribute to achievement of quality in different ways. In the following pages some of these influences at an organizational level are described in six Member States: Denmark, Federal Germany, France, the Netherlands, Sweden, and the UK. The examples show how quality is being achieved through the work of a number of national institutions that make their contributions both individually, and collectively in European and international organizations like CIB, RILEM, ISO and CEN.

The influence that an institution has depends on its history and current standing within industry and government, and on the technological and economic environment within which it operates. The case studies begin with short descriptions of the environment within which an organization works and some of the national factors that influence achievement of construction quality. Some of the institutions, like BRE and CSTB, make their main contribution through research and its application; others, like Federal Germany's DIBt and the Dutch group associated with SBK, through promoting and operating certification services. Two – Denmark's national building agency and Sweden's Boverket – are executive agencies of government with responsibilities for building regulation. BSI is chosen not only because it is one of Europe's leading standards institutions but also because of its pioneering work in the development of quality systems standards.

11.1 DENMARK

11.1.1 Byggestyrelsen: the Danish National Building and Housing Agency

Denmark is one of the smaller Member States, with a population of 5 million, largely urban with nearly a third living in Greater Copenhagen. This fact, despite an efficient local government system, makes for a significant centralizing influence of central government ministries and agencies based in Copenhagen. For construction and construction quality, the major role is played by Boligministeriet (Danish Housing and Construction Ministry), and particularly its National Building and Housing Agency (NBA Bygge- og Bolig-styrelsen).

NBA is of interest for two reasons: the breadth of its responsibilities; and the manner in which the Agency has linked quality assurance, latent defects liabilities and a building defects fund for social housing. Two NBA officials represent Denmark on the EU Standing Committee for Construction (SCC) with technical assistance from SBI (Statens Byggeforskningsinstitut), the Danish National Building Research Institute, which has been designated as Denmark's ETA body and spokesbody on EOTA.

Until the late 1980s the Ministry's executive functions were divided between two agencies: the National Housing Agency for housing legislation, subsidies, support for non-profit-making (social) housing, supervision of mortgage-credit institutions and urban renewal; and the National Building Agency for building legislation, including energy conservation in buildings, development of the Danish building industry, and coordination of public works programmes. In 1988, the two agencies were merged under a single head. Two executive divisions – Internationale kontor and Lejelovs (legal) kontor – have direct international responsibilities, advised by an implementation committee on which industry, the professions and regulators are represented.

NBA has overall responsibility for building regulations, enforcement being a responsibility of local authorities (*kommunatbestyrelsen*), including the City of Copenhagen. Coverage of the technical sections of the 1982 national building regulations differs little from that of 89/106/EEC essential requirements, and, in so far as they apply to the use of traditional materials, the requirement of conformity with Danish Standards, in some cases associated with quality control schemes, replacement of national by European standards presents no difficulty beyond language translation. As European technical approvals begin to be introduced, Denmark's MK-approval system, for which SBI provides technical support, will have to be abandoned. This should provide no difficulty as SBI, a UEAtc member, has been designated the single Danish European technical approval body.

A feature of special interest is the role played in the regulation system by the Danish Society of Engineers, DIF (Dansk Ingeniør-førening). Since 1893, DIF has been responsible for most engineering and construction codes of practice, many of which, following public acceptance, are issued as Danish Standards and quoted in the Danish National Building Regulations. The Society also plays a key role in construction quality and testing through Dansk Selskab for Materialprøvning- og Forskning, the Danish Association for Material Testing.

11.1.2 The Danish Building Defects Fund

Until recently, most housing schemes in Denmark were state-aided, although built and managed by non-profit-making housing associations. These associations, like owners of social housing elsewhere, face higher than normal repair costs from defects arising from the 1970s housing boom, when around 50 000 new dwellings a year were being built – 30 000 with state aid – many using system-building techniques.

In 1985, the Building Defects Fund was set up by the housing ministry, which gave the Fund three tasks:

- to insure the cost of the repair of defects in buildings covered by the Fund, primarily non-profit housing built with state aid after 1986;
- to prevent future occurrence of building defects by stipulating that the construction of all buildings covered by the Fund is quality assured; and
- to build up a database on defects and defect prevention using information gained from the five-year inspections central to the Fund's operation.

All state-aided housing schemes contribute 1% of building cost to the Fund, half being earmarked for the five-year inspection and the rest for defect repairs, subject to a 5% uninsured risk. Only damage that substantially impairs the use of the building for which it was intended – e.g. structural cracking or movement, leaking roof or failure to keep out the weather – is covered. Normal wear and tear, or defects due to lack of maintenance, are not covered. However, failures where untried materials and techniques were used as part of an agreed planned experiment are covered.

Cover is for a period of 20 years after handing over, and carrying out of repair work does not have to wait while defect liabilities are determined. Where a liability can be established, however, the Fund may claim against designer, builder and/or materials suppliers if the defect occurs during the five years after handing over.

The only direct supervision exercised by the Fund is over the five-year mandatory inspection; otherwise, supervision is a matter for the building owner under normal contract practice, and by the local authority under building regulation procedures. However, just before the end of the five-year period, the Fund arranges for a technical inspection by an architect or engineer who is not the original designer. The inspector's job is to assess the condition of the building, record any defects and signs of latent defects, giving reasons for their occurrence. This is done in three stages:

- Stage A: mainly visual inspection;
- Stage B: examination of design drawings, specifications and other contract records; and
- Stage C: where it is difficult to determine the cause and extent of defects, specialist investigation by a laboratory like Dantest or SBI.

A feature of the five-year inspection system is that the nominated inspector is liable for any errors or mistakes he may make in carrying out the inspection during the next five years, giving the building owner cover for a period of 10 years from handing over. By March 1992, the Fund had commissioned about 200 inspections, and the experience gained to date is that the system is effective and the inspection fee fair. The amount that the Fund has itself had to pay out for making good defective work has been small.

At present the scheme only covers state-aided housing, and there is no similar arrangement for private-sector buildings. However, the five-year inspection system is being incorporated in AB 92: *General conditions for works and supplies in the construction sector*. As awarding contracts under AB 92 becomes normal practice, the five-year inspection system is expected to become general. Central government and many municipalities are already introducing the system for non-housing work. In all cases, the building owner will be responsible for arranging the inspection. However, NBA, which is monitoring the scheme, considers that by the time five-year inspections will have become normal practice, there should be more than enough experience to make the system cost-effective.

Under the new arrangements, consultants, suppliers of products used, and contractors are relieved of normal liabilities for defects after the five-year inspection, which means a change in existing liability rules. Up to June 1986 they allowed a consultant's liability to lapse after five years, a contractor's after 20 years and a product supplier's after a year or less. The new five-year liability rule, thought to be a reasonable compromise, only applies to latent defects and not defects that are the result of fraud or gross negligence, the definition of which remains with the courts.

The scheme's aim is to ensure quality in new building during design

and construction, and thus prevent defects rather than allocate responsibilities for the cost of making good damaged work. This is being
achieved through:

- quality assurance during design and execution;
- closer supervision of work during execution;
- rearrangement of acceptance tests;
- closer supervision of maintenance work;
- inspection five years after the building has been completed and
  handed over;
- the setting-up of a building damage fund;
- rationalization of the statutory limitations for the respective liabilities of consultants, suppliers of building products and contractors.

Only the establishment of the Building Damage Fund required legislation; the remaining parts of the package are being introduced through
administrative rules, issued by NBA, or by negotiation between industry bodies.

11.2 FEDERAL GERMANY

> a powerful influence . . . the place of DIN standards which largely
> control the quality of products throughout Federal Germany
> and a selection of which – the ETH (*einheitliche technische
> Baubestimmungen*) – have, through 'legal orders' made by *Land*
> building ministries, the status of mandatory building regulations.
>
> CIOB: *Construction Research and Development:
> A comparative review of four countries*

A justifiable pride in the high standards of industry, preference for
security rather than risking technical innovation, and emphasis on
testing as the route to quality: these are the predominant features
of construction in Federal Germany. Long-established arrangements
to limit building damage through a comprehensive regulation system,
supported by industry standards, and conformity testing and certification, characterize German building and civil engineering. Sixteen
individual *Länder* (State) governments are responsible for most
matters to do with construction, including building ordinances (*Landesbauordnung*, LBO), subsidiary regulations and their administration,
and testing and certification of construction products. *Länder* building
ordinances, based on a model (*Musterbauordnung*, MBO) prepared by
a national committee and maintained by ARGEBAU, a standing
committee of *Länder* building ministers and not by the Federal government, aim to contain risks. By ensuring that work is of a high quality,
the aim is to protect building owners, as well as the general public,
but at a cost.

Specifying requirements very similar to the essential requirements
listed in 89/106/EEC, and placing considerable stress on satisfactory

performance against detailed technical specifications based on German standards, supported by extensive arrangements for testing and certification, the German approach was one of the key influences on the concepts behind the Commission's proposals for the Construction Products Directive.

This approach includes:

- for construction products, stringent production control requirements under which virtually all products undergo some form of technical assessment, usually by a third party, requirements being based on the concept that a specific approval, supported by product marking, was required for any material or component whose performance had not already been confirmed as satisfying the 'technical building rules', i.e. was not covered by an officially recognized standard;
- the distinction in building ordinances between products covered by DINs, new products, and products that because of their intended use must be subject to special controls over quality in manufacture, supported by alternative procedures for verification of conformity to 'technical building rules': manufacturers' declarations; and external (third-party) certification following production control and surveillance, by external control bodies (*Anerkennung Fremd-überwachender Stellen*);
- supervision of testing and factory production control to published standards by RAL, an organization within DIN (the all-German standards body);
- restriction on use of the Ü approval mark to bodies accredited by VMPA (*Verband der Materialprüfungsämter eV – überwachungs-zeichen*);
- a requirement that manufacturers either contract with an officially recognized testing establishment to supervise their factory product control and undertake necessary tests, or join their trade's approved quality control association; and
- long-established links between *Länder* testing establishments, materials testing institutes associated with engineering departments in regional technical universities, and *Länder* construction ministries, with senior academic staff often being *beamter* (established professional civil servants) in charge of the *Land* testing establishment on the university campus, as at Stuttgart.

11.2.1 DIBt, Deutsches Institut für Bautechnik (German Institute for Building Technology)

For products not covered by German standards, certification of fitness for use under a *Land* building ordinance is by means of a *Zulassung* or 'licence'. In 1951, to avoid multiple assessments, arrangements were introduced for mutual recognition of licences issued by individual *Länder*. Principles for assessments were agreed by a national 'committee of experts' against which the validity of a *Zulassung* issued by one *Land* could be extended to the whole of Federal Germany. The arrangement proved cumbersome, and in 1968, by mutual agreement of State and Federal authorities, responsibilities for issuing *Zulassungen* on behalf of *Länder* building ministries were delegated to an all-German body, DIBt (Deutsches Institut für Bautechnik).

DIBt was set up in July 1968 under a special Statute – *Gesetz* – of *Land* Berlin as an *Anstalt des öffentlichen Rechts*. Since the late 1960s, DIBt, like two other all-German organizations – BAM and DIN – has been based in West Berlin. Its work falls to a headquarters administration and four technical divisions: two concerned with civil and structural engineering; one with building services, including fire protection; and one with basic problems including research and the development of standards etc. The work is funded in part by grants from Federal and *Länder* ministries, and in part from fees for its services to industry. It also administers research funds on behalf of its member *Länder* construction ministries.

DIBt has two tasks:

* issuing technical approvals (*Zulassungen*) on behalf of *Länder* ministries for new materials, components and systems not covered by published standards, assessing their performance and properties against the requirements of *Länder* building laws and regulations; and
* giving official recognition to laboratories and quality associations set up by industry to control the quality of manufactured building products.

Since the Single European Market, and especially since 89/106/EEC was adopted, DIBt has provided technical support to the Federal construction and housing ministries in the EU Standing Committee on Construction. It is particularly active in EOTA, its president, Professor Hans G. Meyer, being chairman of the EOTA Technical Board.

Unification has posed problems. First, the new East *Länder* now have the responsibility, like the Western *Länder*, of implementing 89/106/EEC. Second, after a short transitional period, the West German standards have replaced the East German TGL standards. To help industry, the West German standards body has been involved in a comparison of the two systems of standards, as well as the codes of practice and standards issued by the two West German engineering organizations VDE and VDI.

That these events are taking place at a time when harmonized European standards are taking the place of national standards, at least for construction products, gives the coordinating role of DIBt added importance, strengthened by the fact that the institute's president is one of Federal Germany's representatives on the EU Standing Committee for Construction and is chairman of the EOTA technical board.

11.3 FRANCE

> The approach to planning control and building regulation in France is very different to that in UK and the other countries studied. Insurance against decennial responsibilities for building damage under the Civil Code is an important feature of French construction practice.
>
> *Construction research and development: A comparative review of four countries*, CIOB (1988)

In France, close links between the State and its administrators and constructors – architect, builders and civil engineers – date back to the 17th century or earlier. They continue to be a major, if not the dominant, influence on French construction, and on the ways in which construction quality is achieved. The quality of public buildings and civil engineering works is achieved under the direction of a highly trained corps of government civil engineers. Private owners are protected against damage resulting from the work of incompetent designers and builders under Articles 1792 and 2270 of the 1804 Code Civil, amended over the years to take account of changes in client–builder and builder–designer relations, and in recent years strengthened to include the contributions of decennial insurance and independent technical controllers.

11.3.1 The French construction industry and its technical environment

Before examining the very different approach to builder–designer responsibilities, and the way technical control over works is arranged in France compared with, for example, Federal Germany or the UK, it is useful to list a number of features of the administrative and technical environment within which the French construction industry operates.

France is a centralized unitary state, with executive responsibilities for policy matters, including construction, in the hands of cabinet ministers in Paris, advised by highly trained and motivated civil servants. Despite a degree of decentralization, the central state machinery retains a powerful administrative influence throughout France, even though the *mairies*, particularly in Paris and other major centres, have considerable responsibilities in the field of construction, ranging from promoting major public works to granting planning and building permits (*permis de construire*).

The Paris region is the seat of one of Europe's greatest concentrations of state scientific and technological services, whose senior cadre are graduates from a number of highly selective government *Grandes Ecoles*, including *l'Ecole des Ponts et Chaussées* for civil engineers. Consequently, central government, state and private industry are served by civil engineers who have had a broad theoretical training in technical and scientific principles, followed by a period of practical experience in management of public works in the field before a policy-making career that may include directing work in organizations like AFNOR (the national standards organization), technical control offices like SOCOTEC, newer agencies like *Agence pour la Prévention des Désordres et l'Amélioration de la Qualité de la Construction (Qualité Construction)*, and CSTB, as well as in national industries.

MELT, the ministry of public facilities (*équipement*), housing (*logement*) and transport, with headquarters in the Grande Arche, La Défense, does not only have responsibilities for construction. It is the housing ministry, and also has a major interest in all forms of transport as well as some aspects of physical planning (*aménagement du territoire*). Interventionist by tradition, it is served by executive units at regional and *département* levels. CSTB (Centre Scientifique et Technique du Bâtiment), a public establishment under the tutelage of the ministry's Department of Housing and Construction, has a major responsibility for the achievement of construction quality.

France has a long tradition of carrying out very large construction projects in the field of public architecture, such as the Cité des Sciences et de l'Industrie at La Villette, the Grande Louvre, the Musée d'Orsay in Paris, and the new Bibliothèque, through special agencies – *établissements publics* – set up by presidential decree to ensure a separation between the *maîtrise d'ouvrage* or client organization and the *maîtrise d'oeuvre* or project team.

Social housing, whether rental, cooperative or subsidized privately owned and occupied, is a responsibility of HLM organizations, which can call for assistance on management, training and technical matters from a central Union. The HLM (Habitations à Loyer Modèré – 'housing at reasonable rents') movement approaches social housing in a flexible way and allows partnerships between a local HLM housing *organisme*, regional and national agencies, and private enterprise advised by experienced central Union services under the general supervision of the housing ministry.

French contractors, led by Bouygues, are among the biggest in Europe, some being linked with major urban services concessionaires like Compagnie Générale des Eaux and Lyonnaises des Eaux. Although the contractors' technical organization, UTI, is not as strong as it was in the 1970s, the industry still has a large training centre and research and testing laboratory at St-Rémy-les-Chevreuse outside Paris.

However, France lacks the independent and possibly more pragmatic approach to construction quality that UK professional institutions like the ICE, RIBA and RICS provide; nor does it have the strength of an independent academic engineering fraternity as found in the institutes of Germany's technical universities. Additionally, while firms like Bouygues and Compagnie Générale des Eaux employ engineers of high quality, many trained in one of the *grandes écoles*, France lacks the independence of the British, possibly multi-disciplinary, professional consultant.

Codified under Napoleon I, laws, adopted in the Assembly, usually set out principles, brought into administrative operation through *décrets* – decrees issued by the prime minister after consultation with the ministers concerned and the *Conseil d'Etat* (constitutional court); and, for detailed application, through *arrêtes* – orders issued by the responsible minister. To Anglo-Saxons unfamiliar with Roman law, the process may seem bureaucratic. But, for those familiar with the system, it has three merits:

- Documents are logically arranged and clearly written, with technical matters explained, where appropriate, using mathematical formulae.
- Legislation is codified and, from time to time, made readily available through bookshops to practitioners and the general public in updated codes, such as the *Code Civil*.

Extracts of relevant legal documents are usually published in full in a looseleaf supplement to the weekly construction journal *Le Moniteur*, and later bound in convenient collections.

The result, at first sight bureaucratic, is easier for a lay person familiar with the system and with the French language to understand than English legislation, and seems to depend less on interpretation by lawyers.

Under the Civil Code, updated in 1978, there is a clear definition of legal liabilities for damages arising in the first ten years after completion of works. Owners and constructors insure, and there is near-mandatory control by licensed technical offices, some of which, like the largest, SOCOTEC, operate strong technical services based on research and practical experience.

France, one of the original signatories to the Treaty of Rome, had and continues to have a major influence on European Union legislation and institutions, an example being the adoption in 89/106/EEC of the alternative ETA route to CE marking based on the *agrément* concept, initiated by CSTB in the 1950s and introduced into the UK in the 1960s.

11.3.2 The influence of Civil Code responsibilities and construction quality

The French approach to regulating building quality, based historically on the protection of owners against building damage under a Civil Code, which makes constructors – designers and builders – responsible for any major damage for ten years, merits special mention. It is in marked contrast with that of Germany, based on a 19th century Prussian concept of orderliness, or that of the UK, the Netherlands and Scandinavia, based on modern technical regulations, supported by advisory documents and national standards, and administered by local authorities (see Box 11.1).

The 1804 Civil Code has for nearly two centuries had a dominant

BOX 11.1   THE FRENCH AND GERMAN SYSTEMS COMPARED

The German System

'More control by authorities:
less responsibilities by constructors'

Strict control during the construction process:
* stringent regulations
  based on full national standards (DINs)
* checking of designs before issue of building permit where necessary by an independent licensed structural engineer (*prüfingenieur*)
* products either tested against DINs and subject to factory production control, or approved by DIBt, the German Institute for Building Technology, Berlin
* technical site inspection, with much attention given to final inspection to uncover deficiencies

The French System

'Less control by authorities
more responsibilities for constructors'

No technical control by authorities during design or construction:

* building permits issued against signature of an architect that design meets planning, health and fire safety regulations with possibility that compliance checked by government engineer in first three years after completion
* clearly stated guarantees for soundness of and fitness for intended used within specified time limits
* requirement for building owner's damage insurance and constructors' responsibility insurance
* mandatory requirement for independent technical control for high-risk situations and preference for building owner to employ otherwise
* importance placed on *réception* or practical completion as start of damage insurance

influence on French construction practice. To protect building owners, it required designer and builders – the constructors – to guarantee the soundness of structure and weathershield for ten years. After this period, the 'professionals' were discharged from their responsibilities.

Since the 1800s, Civil Code clauses and related legal texts have been amended and extended to reflect technical changes, and changes in forms of ownership and responsibilities within the building industry. For minor works like building equipment a shorter guarantee period was introduced, as was the concept of 'fitness for intended use'. In 1967 the fact that others than architects and builders might have responsibilities for building performance was recognized. There were further changes in 1972, when the insurance industry's technical organization was strengthened, and procedures were introduced for acceptance of non-traditional materials and systems not presenting abnormal risks.

There were more fundamental changes in 1978 under the Spinetta Law on 'Liability and insurance in the field of construction'. With limited exceptions, building owners have to take out a damage insurance – *assurance dommage* – policy to ensure that major repair work is not held up while contractors and designers argue over responsibilities and costs.

Architects and builders are required to guarantee conformity of design and construction with regulation requirements. Under the Architects' Code, the architect has to carry responsibility insurance for design matters, checking usually being by the licensed technical controller employed by the building owner. Where structural and fire safety risks are high, employment of a technical controller, who has to be licensed by a government commission, is mandatory.

The 1978 system has a number of interesting features:

- Technical control is operated by independent licensed offices chosen and paid by the building owner, to whom they are legally responsible, and not, as before 1978, employed by the insurance industry.
- By making employment of a technical controller mandatory for certain types of work, it brought controllers into the regulatory system, but under different conditions from those in Germany, or the Netherlands, Scandinavia and the UK.
- Not only has a building owner a number of licensed offices to choose from, he can also commission from the office chosen services beyond mandatory requirements.
- Larger control offices have branches throughout the country and, in some cases, outside metropolitan France. Their wide coverage enables them to build up large technical databases.
- With clearly defined changes in responsibilities for insurance when works are handed over, importance in French jurisprudence is placed on *la réception* – sometimes inaccurately equated to 'practical completion'.
- Technical control is mandatory in major situations where there is a potential risk to occupants, like structural collapse and building fires.
- Local *mairies* receive, process and publish planning and building permits (*permis de construire*), but have little or no direct responsibility for ensuring that technical requirements of the Building Code are satisfied, this being a matter for the technical office as the building owner's agent, who may also be commissioned to undertake supplementary services such as avoidance of damage

to adjoining property and existing structures retained on a site, protection of contents against fire damage, and protection against earthquake damage.

- Where a building owner contracts with a technical controller for services additional to mandatory controls, there are reductions in damage insurance premiums.
- Technical control services covering testing and operation of technical installations like heating, air conditioning, plumbing services, electrical installations and lifts can be arranged by building owners with larger offices.

11.3.3 The French building research centre: CSTB, Centre Scientifique et Technique du Bâtiment

Founder member of CIB and UEAtc (European Union of Agrément) and active member of ENBRI, CSTB is one of Europe's leading centres of building technology. It initiated the *agrément* system for assessment of fitness for use of new products and techniques, and by providing the secretariat of UEAtc has done much to promote the *agrément* concept throughout Europe. It is now one of two ETA bodies designated by the French Government (the other being SETRA, the road construction technical service), and its previous president, M. Pierre Chemillier, is France's principal representative on the EU Standing Committee for Construction, being the first chairman of EOTA .

Established in 1947 as the Centre Technique du Bâtiment et des Travaux Publics, CSTB opened an experimental site at Champs-sur-Marne on the outskirts of Paris, moving its headquarters, a specially designed office block with lecture rooms etc., to the west end of Paris in 1951. At the start, some of CSTB's testing activities clashed with those of the civil engineering laboratory of FNB, the contractors' federation, and when in 1953 it was given a new statute as a public establishment, civil engineering was excluded from its terms of reference. CSTB was then able to centre its work on building and housing.

A few years later it had the task of preparing national building regulations (*règlement général de la construction*), mainly for health and sanitation in housing. The DTU series of technical specifications – to be equated with BS Codes of Practice – followed. In 1958, CSTB became responsible for approval of new products and systems for use in state-financed housing, operating a system of assessment and approval for specific uses of products and techniques not covered by French *normes* (standards). Named *agrément*, it gave a yes-or-no decision on the fitness of new products or techniques. In 1969, following demand from manufacturers for greater involvement, final responsibility for issuing renamed *Avis technique* certificates was transferred to an inter-ministerial commission, chaired by a senior civil engineer in the construction ministry, on which industry, technical control offices and HLM (the social housing movement) were represented. CSTB continued to provide the secretariat, its scientific and technical staff undertaking assessment and test work, helped by some 15 specialist technical groups.

Avis techniques serve as independent sources of information on likely performance and fitness for use of products and techniques; and as a way of enabling the insurance industry covering constructors' decennial responsibilities to assess risks involved in the use of new products and techniques. The industry's technical committee from time to time, publishes lists of approved products given *Avis techniques*, considered to present normal insurance risks if used in accordance with the terms of the certificate and any further limitations set by the committee. More recently, as an aid to innovation and a preliminary step towards an *Avis technique*, CSTB initiated ATEx (technical experimentation assessment), which helps experimental application of a new technique by limited assessment directed at safety, feasibility and risk of defects.

CSTB is an *établishment public de caractére industriel et commercial*, which allows it as a public body to manage a budget agreed with the construction and finance ministries, adding to 'grant-in-aid' income revenue earned from research contracts, selling services and publications, running training courses etc. Only the president, appointed by decree of the prime minister on advice of the construction minister, and a few of the senior staff are *fonctionnaires* – permanent civil servants attached from the construction ministry.

CSTB differs from Germany's DIBt in two respects. First, an *Avis technique*, although much used in state-funded projects and in damage insurance, is not mandatory like a *Zulassung*. Second, as CSTB is under the tutelage of a single construction ministry, it has significantly broader research responsibilities, using its own laboratories. Again, although CSTB could be said to equate to BRE and BBA, there are some significant differences:

- CSTB undertakes *agrément* work, undertaken in the UK by BBA, the British Board of Agrément.
- CSTB does not have the benefit of the BRE Fire Research Station's unique Cardington 'burn-hall' facility.
- Most work on timber and wood products falls to Centre technique du Bois et l'Ameublement.
- Its work in civil engineering and public works is limited.

CSTB's main research laboratory is at Marne-La-Valée, a new town east of the capital, where much of the work in building technology, structural engineering and fire research is undertaken. Other centres are at Nantes, on the Atlantic coast (climatology, aerodynamics, ventilation and lighting); Grenoble, near the Dauphine Alps (acoustics, sound insulation, and the properties and performance of building materials); and Sophia-Antipolis, a 'science city' near Nice on the Mediterranean coast (solar energy research and application of information technology to CAD, artificial intelligence and robotics).

In France, CSTB is the main source of construction information. One of its early responsibilities was REEF, an encyclopaedic looseleaf construction reference work. The 6th edition comprises over 14 000 pages in 18 volumes, which include:

- a complete set of DTUs (*documents techniques unifiés* – roughly to be equated to design and workmanship codes of practice);
- *cahiers des charges* (standard forms of contract);
- French national standards relating to building;
- building legislation and official technical regulations;
- *règles de calcul* (design codes) with examples of solutions;
- a Science du Bâtiment series; and
- Avis technique certificates.

CSTB's importance as a central source of construction information is demonstrated, not only by the range of its publications listed below, and by a recent responsibility for distributing the publications of the housing ministry's Plan Construction et Architecture, plus Syncodés and other publications of the Agence Qualité Construction, but also by its initiatives in electronic publishing. REEF is available by subscription on CD-ROM, and CSTB provides one MINITEL service, 36-16 CSTB, which gives access to a number of databases, including ACERFEU on fire resistance certification. It also collaborated with the contractors' organization ITB in a second service, 36-17 BATI-BASE.

Its established publications, listed in the CSTB annual catalogue, include:

- *Cahiers de CSTB*: a collection of research papers, full texts of new DTUs, UEAtc directives and papers in the EPEBat series, which appears ten times a year, having an average of 250 pages, with an optional supplement of new *Avis techniques*;
- *CSTB Magazine: Une Reference pour le Bâtiment*, giving technical information for architects, engineers and builders on results of both CSTB work and technical developments generally;
- *Collection Artisanat*, a series sponsored by the Ministry of Commerce and Small Firms (*Ministère du Commerce et de l'Artisanat*), which is devoted to commentaries on regulations and codes of practice (DTU) and technical guides of special interest to the small builder.

11.3.4 Agence Qualité Construction

Separate from CSTB but closely linked under the tutelage of MELT, the housing ministry, is the Agence Qualité Construction, a non-profit-making agency set up in 1982 to implement one of the Spinetta law recommendations. Working with construction damage insurers, through a network of 300 experts, the agency is building up a database of construction defects and failures, which are analysed and commented on through a bimonthly journal, *Syncodés*, and a series of technical guides.

11.4 THE NETHERLANDS

In the Netherlands, attitudes to achievement of construction quality are more liberal. They are based on a complex but rational approach involving central and local authorities and industry in a many-sided network, which is of interest for the number of organizations involved (Figure 11.1).

Unlike the situation in France and the UK, NNI (Nederlands Normalisatie-instituut), the Dutch standards body, is not also a certification body, nor does it operate a testing laboratory. Its influence on the certification process is through membership of RvC (Raad voor Certificatie), the Dutch Council for Certification, HCb (Harmonization Commission for Building) and, internationally, through membership of

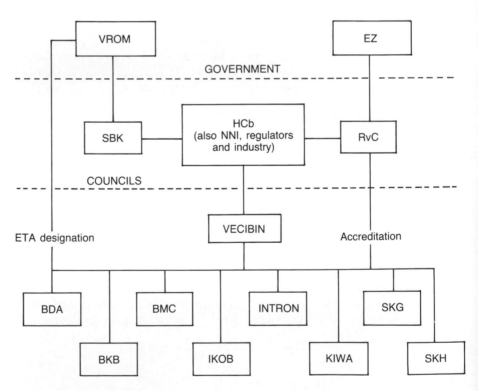

Figure 11.1 The Netherlands construction quality network. VROM: Ministry of Housing etc. EZ: Ministry of Economic Affairs. SBK: Foundation for Building Quality. RvC: Dutch Council for Certification. NNI: Dutch Standards Institute. HCb: Harmonization Commission (building). VECIBIN: Association of Certification Bodies within the Construction Industry. BDA, BKB, BMC, IKOB, INTRON, KIWA, SKG, SKH: Dutch certification bodies (see Part Two, Chapter 15 for names and addresses of these organizations).

ISO and CEN. Furthermore, no single Dutch body had the monopoly of issuing *agrément*-type certificates, and eight bodies have been recognized by the Dutch government to issue European technical approvals.

Accreditation of certification bodies is by an independent, state-supported council, RvC. Unlike the UK, however, there are no separate organizational arrangements for testing laboratories; instead their certification is through STERLAB (Nederlandse Stichting voor de Erkenning van Laboratorie), a body accredited by RvC. There are to be similar arrangements for inspection services. RvC was set up in July 1981 on the initiative of the Ministry of Economic Affairs as an independent non-profit-making organization (*stichting*). The Council has had strong initial financial support from the Dutch Government through a grant towards operational expenses on a decreasing scale. Its headquarters is at Driebergen, a small town between Utrecht and Arnhem. It has accredited over 20 certification bodies, including a number from other Member States, and is negotiating an agreement with the UK's NACCB for mutual acceptance of assessment reports, on the basis of which accreditation is granted.

11.4.1 SBK, Stichting Bouwkwaliteit (Foundation for Building Quality)

The special needs of construction are recognized in the setting up of SBK (Stichting Bouwkwaliteit), the Foundation for Building Quality, an independent foundation under the sponsorship of VROM (Ministerie van Volkshuisvesting, Ruimtelijke Ordening en Milieubeheer), the Ministry of Housing, Physical Planning and Environment. SBK manages the most widely used construction quality mark scheme – the KOMO mark – but does not itself carry out testing, inspection or certification. This activity is the responsibility of a number of independent organizations, of which KIWA (Keuringsinstituut voor Waterleidingaretikele), Approval Institute for Waterservices Products, is the largest. Three kinds of document are issued:

- *certificaats*: certificates of conformity to a standard;
- *attests*: certificates of fitness for use; and
- joint *attest-met-certificaat* certificates.

This has meant that, under the Dutch arrangements to implement 89/106/EEC, instead of entrusting the issue of ETAs to a single organization, as in Denmark, Germany and the UK, eight possibly competing bodies have been designated as ETA bodies by the Dutch government All are members of VECIBN (Vereniging van Certificatie-Instellingen in de Bouw in Nederland), the Association of Certification Bodies for Building Components and Materials, which publishes a consolidated annual list of certificates issued.

SBK has a coordinating function on matters relating to quality of construction products, providing the main forum for discussion within the Dutch construction industry of EU quality issues, its quarterly publication *Bouwkwaliteit-Nieuws* giving detailed information on national and European events. It is the Dutch member of UEAtc, and is the Dutch spokesbody on EOTA and participates in working groups preparing ETA guidelines.

An interesting feature of the Dutch arrangements is the setting up of HCb (Harmonization Commission for Building) under the joint administration of RvC and SBK, which has the task of coordinating criteria for conformity, fitness for use and quality system certification. HCb's membership includes VECIBIN (the Association of Certification Bodies within the Construction Industry), NNI (the Dutch Standards Institute) and representatives of industry and regulatory bodies. SBK provides the secretariat.

11.4.2 Building quality: arrangements at government level

Arrangements for implementing 89/10/EEC are taking place in parallel with the adoption of performance-based national building regulations, which will replace a previous system of building control through municipal byelaws. VROM, the Dutch ministry of housing, planning and construction, through its Directoraat-Generaal van de Volkshuisvesting (Directorate-General for Housing) and Directorate for Research and Quality Assurance is responsible for the implementation of 89/106/EEC with EZ (Ministerie van Economische Zaken), the economic affairs ministry. Officials from the two ministries represent the Netherlands on the EU Standing Committee for Construction, their permanent technical expert on the Committee being Ir E.G.C. Coppens, a professional civil engineer formerly in VROM and now director of SBK.

On EU construction matters, VROM has set up a consultative machinery with the professions, industry and regulators, and a smaller official committee (Bestuur van het EG –Beraad voor de Bouw), aided by a three-person secretariat. Following adoption of a new Housing Act (*Woningwet*) by the Dutch Parliament, VROM has been preparing performance-based national building regulations, for which there is a separate consultative machinery, technical drafting being entrusted to TNO-Bouw, the building materials and structures research institute. Close links have been forged between the consultative machinery for regulations and implementation of 89/106/EEC, and the regulations can be regarded as one of the first national regulations so influenced.

11.5 SWEDEN

As a member of EFTA, Sweden has been participating actively in European standards-making in CEN. In anticipation of Sweden joining the European Union, Boverket, the central government agency with responsibilities for product certification and factory production control as well as building regulation and housing and physical matters, has joined UEAtc and is likely to be designated the Swedish ETA body. The Nordic countries have for some time cooperated in the development of mutual arrangements for product certification and factory production control. It is, therefore, possible that if the Commission had investigated the Nordic arrangements for a single market in construction products, European industry might have been subject to a simpler, possibly less bureaucratic system than that adopted in 89/106/EEC.

Other noteworthy features of Swedish arrangements for achievement of construction quality are:

- the delegation of ministerial policies on social and technical matters to professionally staffed executive agencies (*ämbetsverk*), which carry out tasks assigned under legislation within a financial programme;
- for construction matters, the delegation of central responsibilities for building regulation, and associated arrangements for certification and quality assurance, to Boverket (the National Board of Housing, Building and Physical Planning), or to one of two other construction agencies, KBS (National Board of Public Buildings) and the National Road Administration;
- administration of building regulations by local district authorities, helped by a well-presented system of approved documentation;
- regulations – the Swedish Building Code – in which requirements, expressed in functional or performance terms, are supported by examples of solutions, recommendations and advisory information either within the Code or related approved documents;
- an associated system of type or general approvals (*typgodkännanden*), issued by Boverket, supported by testing and factory production control, in the operation of which regulators and industry participate, and backed by informative marking;
- long-established arrangements for technical cooperation between the four Nordic countries – Denmark, Finland, Norway and Sweden – on building regulation, certification and approval of products, and testing practices.

Arrangements for ten-year warranties for protection against major building defects in state-funded housing through AB Bostadsgaranti, a company owned jointly by government and the Swedish contractors' federation, Svenska Byggnadsentreprenörföreningen, are now mandatory. For non-residential building, there is no similar arrangement, liabilities and guarantee periods being covered by the General Conditions of Contract AB72 and ABT74, and for architects and consulting engineers ABK 76. Under the 1987 Planning and Building Act, following a decision of the Swedish Supreme Court, a local authority giving building permits may be liable for defects where it has a particular knowledge of local conditions not known to the building owner.

Boverket, the National Board of Housing, Building and Physical Planning, is housed in the country's oldest remaining naval barracks at Karlskrona on the Baltic coast, some 600 km south of Stockholm. One of Sweden's newer government agencies, it was formed in July 1988 by an amalgamation of the National Housing Board and the National Board of Physical Planning and Building (Statens Planverk), which, since 1968, had been the agency responsible for building regulation and the central authority for building product approvals, as well as the central agency for physical planning.

The shift from a largely regulatory to a more knowledge-based role in housing and building matters reflects a move away from the rather isolated and neutral approach to developments in western Europe that dominated the policies of Swedish governments in the 1960–1970 years. It has meant that Sweden's powerful technological experience in construction matters is now available in a wider European context through CEN and Boverket's membership of UEAtc. At the same time, Boverket's links with other Nordic countries on building regulation matters continue in NKB (the Nordic Committee on Building Regulations).

11.6 THE UNITED KINGDOM

Architects always have worried about quality, it is the backbone of their professionalism, but with the increasing complexity of buildings, both in the range of clients' requirements and in the increasing technologies that are available to satisfy needs, it is vital that far more effort and resources are devoted at the inception and feasibility stage to determining what is required and to examining the alternative options available. Having done so, then, and only then, is it possible to draw up a firm brief . . . in which the required quality standard is identified by the client.

M J Veal: Presentation on behalf of the Royal Institute of British Architects, National Quality Assurance Forum for Construction (1988)

Architects are but one of a number of UK professional bodies with an interest in construction quality and its achievement, although as agents of building owners they have a special interest with their colleagues in other professional bodies. A feature of UK construction is the number of these bodies and their mutual interest in quality and quality assurance.

Cooperation takes many forms: for many through participation in standards-making work; for some in serving on the boards of organizations like the British Board of Agrément. They are supported by the Construction Sponsorship Directorate of the Department of the Environment, and in turn with industry support the Department's representatives in the EU Commission's Standing Committee for

Construction. It is useful for these reasons to look at the way DoE's construction activities have recently been strengthened by the establishment of five Divisions: Construction Export Promotion; Construction Industry Sponsorship; Construction Products and Materials Sponsorship; Market Intelligence; and Innovation and Research Management.

This last Division works with industry and the professions to develop and foster a strategy for research, innovation and technical knowledge, managing a DoE programme of some £25 million directed, among much else, to supporting the UK's contributions to the development of European codes and standards, mainly through the British Standards Institution, the UK's national standards body. In this activity it draws considerably on the technical and scientific support of DoE's Building Research Establishment.

11.6.1 Building Research Establishment (BRE)

Since April 1990 BRE has been an Executive Agency of the Department of the Environment. It is the principal UK organization undertaking research into building and construction, and into the prevention and control of fire. Its main role is to advise and carry out research for Government, principally DoE, on technical aspects of buildings and other forms of construction, all aspects of fire, and environmental issues relating to buildings. BRE also manages a major programme of information transfer for the Energy Efficiency Office.

BRE's research ranges from studies of the basic properties of materials to investigations into the performance of complete buildings. It carries out the work through six management groups covering, respectively:

- materials;
- construction practice, economics and application;
- fire;
- environment and energy;
- geotechnics and structures; and
- financial planning and resources.

Its fire research group, at present located at Borehamwood, Hertfordshire, will soon join the main establishment at Garston, near Watford, from where it will operate one of Europe's largest fire and structures test facilities at Cardington, Bedfordshire.

BRE provides the technical basis for the Building Regulations and many British codes and standards. It supports the UK Government and industry in the implementation of the Single European Market, particularly in providing technical advice on the implementation of 89/106/EEC, where it has made a major contribution to preparation of the Interpretative Documents, as well as working on mandates issued by the Commission for the preparation of European standards and Eurocodes. For DoE, BRE's international division has the task of developing and maintaining a register of organizations competent to test construction products for CE marking, with details of the tests that must be performed.

For many years BRE has been an active member of CIB and RILEM; it is a leading member of ENBRI (European Network of Building Research Institutes}, and has been designated by the European Commission as an Organization for the Promotion of Energy Technologies.

BRE's experience in the application of scientific and technological research through technical publications, among which the Building Research Digests are the best known by practitioners, goes back many years. It is the largest single publisher of technical information to the construction industry in the UK. The BRE Bookshop holds publications of the Property Services Agency and the Energy Efficiency Office, as well as its own publications: Good Building Guides, Digests, Information Papers, Overseas Information Papers, Building Notes and Technical Reports. In the early 1970s, BRE carried out a large-scale review of the procedures in various European countries for the approval of new and established products used in construction. At the conclusion of the study a summary report was published as BRE Digests 166/167: *European product-approval procedures*. Recently, as background to the implementation of 89/106/EEC, it published BRE Digest 376: *European legislation and standardization*.

11.6.2 British Standards Institution (BSI)

Unlike BRE, BSI, as the national standards body, covers the whole field of industry and its technologies. It claims to be the first national standards body in the world, having been formed in 1901 as the Engineering Standards Committee. The national importance of this Committee's work was confirmed when, as the British Standards Institute, it received a Royal Charter in 1929, giving it the tasks of:

- coordinating efforts of producers and users for the improvement, standardization and simplification of engineering and industrial materials, so as to simplify production and distribution, and to eliminate the national waste of time and material involved in the production of an unnecessary variety of patterns and prices of articles for one and the same purpose; and
- setting standards of quality and dimensions, and preparing and promoting the general adoption of British Standards specifications and schedules in connection therewith.

As an independent organization operating under Royal Charter, BSI is the UK's national standards body, and a member of ISO and CEN. Since the publication in 1982 of the Government White Paper *Standards, quality and international competitiveness*, much has been done to strengthen BSI's organization and management. There has been closer cooperation between Government and BSI in the development of British Standards suitable for regulatory purposes and/or public purchasing.

There is an associated commitment to make greater use of standards, where appropriate, in Government's regulatory functions, and to place greater emphasis in public purchasing on linking requirements to existing standards rather than to technical specifications of public purchasers. Certification schemes were encouraged with the introduction of unified arrangements for laboratory accreditation through NAMAS and for certification bodies through NACCB.

BSI work is centred around five activities: BSI Standards, including BSI Sales and Customer Services, Information Services and Technical Help for Exporters; BSI Publications and Information Services; BSI Quality Assurance; BSI Testing; BSI International Training; and BSI Product Certification. Some 27 500 organizations, as subscribing members, provide about 10% of BSI's annual income, which in 1991/1992 was £100 million. To help meet the challenge of the 21st century, BSI has relocated its head office and BSI Standards in new offices at Chiswick, equidistant between Central London and London (Heathrow) Airport.

Standards-making remains BSI's main activity. Standards underpin the law, and are an essential component of New Approach European legislation. Anyone claiming unilaterally that a product or service is in conformity with a British Standard assumes a legal responsibility for the accuracy of the claim, which, if shown to be demonstrably false or misleading, is liable to prosecution under the Trades Description Act. The scale of BSI Standards work is demonstrated in that it provides the technical secretariats of nearly 3000 UK and around 200 European and international committees and working groups.

With most standards-making being directed at international and European work, the UK no longer controls the content and volume of a growing number of standards, but, through a strengthened BSI, it can have a key influencing role in the outcome through a number of related actions:

- stimulating better international and European standards-making procedures: for example, avoiding duplication of effort at European and international levels, making greater use of appropriate national standards as the basis of European standards, and ensuring that UK delegates are of the right calibre and seniority, and adequately trained and briefed;
- promoting greater use of standards in technical legislation, certification and public procurement;
- safeguarding consumer interests, a responsibility delegated to the Standards Board's Consumer Policy Committee;
- back-up to UK participation in the preparation of standards that best serve UK industry, trade, consumers and government legislation; and

- establishing a standards information base best suited to the needs of UK industry and other users.

The change in standards-making from national to international, and especially European standards, is reflected in new arrangements for building and civil engineering standards where, under B/- (Technical Sector Board for Building and Civil Engineering), a structure and method of working and structure reflecting that of work in CEN have been created aimed at:

- matching staff and committee member resources with work to be done at national, European and international levels; and
- striking a balance between service to industry and professions and rapid decision-making, small committees, short lines of communication and minimum paper work.

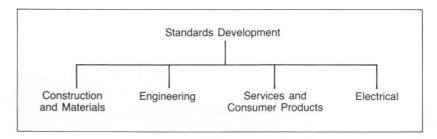

Figure 11.2 BSI standards-making organization.

Standards-making and maintenance of the existing body of construction standards fall to a three-tier system of committees, subcommittees and working groups. The first consists of members nominated by bodies needing the standards and having responsibilities for planning, programming and coordination and organizing the necessary standards-making work; the second has responsibilities for preparation of technical content of the standard; and the third undertakes limited allocated tasks within the standards-making work.

As standards-making proceeds, to ensure coordination of related technical committees' work, B/– may set up a panel to deal with a complex area of work. One example is B/–/2, the Conformity and Quality Panel for Building Products, which has the task of advising B/– on requirements for verification and attestation for national, international and European standards.

BSI is a publisher of some importance, publishing, in addition to standards and related documents: an 800-page *Annual Catalogue*, in which British and corresponding international standards are listed; a range of handbooks and other guides; the monthly *BSI Update*, in which new work is reported as well as lists of updated and new British and international standards; and periodical bulletins such as *Technical Export News*, *Environmental News* and *Standards News International*.

Construction presents special problems for standards publication, for a number of reasons:

- the range of subjects covered, especially with the growing importance of building environmental and security services and fire protection and safety;
- the long-established use in design and sitework of codes of practice, many of which, but not all, are a BSI Standards responsibility;
- the width of technical sophistication among users of standards, ranging from testing scientists and engineers in specialities like geotechnics to the generalist architect and builder, many of whom are in small or one-person businesses; and
- the widespread use by reference of standards and codes of practice in regulations and contract documents.

To meet such varying needs, BSI has undertaken a range of initiatives. For constructors, detailed summaries of over 1500 standards documents are collected in a four-volume looseleaf *BSI Handbook 3*. Between 1982 and 1990, a series of guides on the preparation of British Standards for building and civil engineering have been issued as PD 6501, which covers: aims, content and application; presentation; needs of users of building regulations; and the principles involved in preparation of performance specifications.

Technical Help for Export (THE) is a service for exporters, which provides a consultancy, information and translation service aimed to help British firms with information and advice on the requirements of European and other technical regulations, standards and testing procedures. Its translations, available from BSI Publications, include a number of French and German codes and regulations. THE issues a quarterly bulletin, *Technical Export News*, in which developments in the Single European Market and elsewhere are reported, and recent translations are listed.

FURTHER READING

Denmark

Ann Alderson, *Denmark: Product approval, quality control and building control procedures*, Building Research Establishment Report (HMSO, London, 1974)
An earlier and more detailed study.

G. Atkinson, *European product certification schemes: Denmark*, Building Technical File No 25 (April 1989)
Gives background to Danish construction; Nordic cooperation; building regulations and current arrangements for product approvals and certification with reference to the 'shall' (compulsory) and 'can' (voluntary) systems; conformity certification to DS Standards; and accreditation of testing laboratories. The principal organizations

concerned with building regulation, and building products approvals, certification, testing and quality control are listed.

Danish National Building and Housing Agency, *Quality assurance and liability in building*, Report of a CIB meeting in Copenhagen April 1992, prepared by the Agency, from whom up-to-date information may be available.

Federal Germany

G. Atkinson, *European product certification schemes: Federal Germany*, Building Technical File No 22 (July 1988) and No 29 (April 1990)
Reviews the influences of State and Federal legislation, and the roles of DIBt and other organizations.

W.D. Biggs, M. Betts and M.J. Cottle, *The West German construction industry: A guide for UK professionals*, CIRIA Special Publication 68 (1990)
Gives the national background to the German construction industry, and opportunities for British and German consultants and contractors. Appendices: miscellaneous construction statistics; annotated bibliography; glossary of German terms; glossary of abbreviations; directory of German organizations.

BSRIA, *Key Standards for the Building Services – Germany* (BSRIA Euro Centre, Bracknell RG12 7AH, 1991)
Lists 633 standards from over 50 000 entries in 1991 DIN Katalog für technische Regelen relating to products and installations in areas of heating, ventilation, air conditioning, environmental controls and plumbing.

Elaine Cesselin, *La réglementation de la construction en République Fédérale d'Allemagne* (CSTB/Norex, November 1987)
A French study of the German system of building control.

Evelyn Cibula, *Building control in West Germany*, BRE Current Paper CP10/70 (1970)
An earlier study of the legal and administrative background, procedures and enforcement, standards and approval schemes, technical content of regulations.

A. Doran, *Craft Enterprises in Britain and Germany: A sector study* (Anglo-German Foundation, London, 1985)
Describes the well-organized system of small *Handwerk* firms found in Germany, who are able to maintain a high level of skills through the statutory backing given to trade definitions and training standards and the obligation of school leavers to undertake vocational training.

G. Fuhrmann (BAM) and H.G. Meyer (DIBt), *Agrément und Zalussung*, UEAtc Information 2 (May 1980)

Hans Gerd Meyer, The evolution of building techniques in the Federal Republic of Germany, *First Symposium of the European Building Industry*, Lyons (November 1986)
A review of the high requirements of German construction, by the head of DIBt.

Ernst Neufert, *Bauentwerfslehre. Handbuch für Baufachmann, Bauherrn, Lehrenden und Lernenden Vieweg*, 33rd edn (Weisbaden, 1991)
A standard German design and construction textbook.

P. Patel and K. Pavett, *Technological activities in the Federal Republic of Germany and the UK* (Science Policy Research Unit, Sussex University, 1988)

Wenderhorst/Spruck, *Baustoffkunde* (Curt R. Vincent Verlag, Hannover, 1986)
A typical example of German encyclopedic textbooks on construction materials.

France

G. Atkinson, *European product certification schemes: France*, Building Technical File No 23 (July 1988)
Gives background to French construction, DTUs (specifications and codes of practice), *Avis techniques* (technical assessment of products and techniques), accredited certification bodies; Marque-NF. Technical control arrangements, construction product databases.

G. Atkinson, *French system of technical control: its merits and limitations*, Building Technical File No 27 (October 1989)
The Spinetta reforms; types of services; technical control; the AuxIRBat study; quality assurance; major technical controllers.

G. Atkinson, *Technical building control in France: recent changes to the contractors' basic Civil Code liability insurance policy and consequent use of Avis technique successor to Agrément certificates*, Building Research Station Current Paper 46/74 (1974)
An earlier study.

Pierre Chemillier (CSTB), Performances/ l'Avis technique se met au gout du jour, *Le Moniteur* (20 December 1985)
With the issue of the 5000th *Avis technique*, plans for changes in their form directed to a performance approach are described.

CREPAUC, *Les marchés privés de travaux: Documents types et commentaires avec texte intégral de la norme NF P 03-001* (Editions du Moniteur, 1986)
Cahier des clauses administratives particulières (CCAP) includes: *réception des ouvrages; assurances: autres documents types; demande de réception; procès-verbal de réception; constat de levées des réserves*; the French code for private works contracts.

André Grelon, *Les Ingénieurs de la Crise* (Editions EHESS, Paris, 1986)
Includes information about the training of civil and structural engineers in Federal Germany and France, as well as other countries.

Guide Pratique: Certification des Produits de Construction, Qualité Construction, 2nd edn (1991)

Jens Knocke, *Prevention of building damage – the French way*, National Swedish Institute for Building Research, SB: 49 (April 1992)

Pierre Maurin, *l'Evolution de l'Assurance Construction* (Editions L'Assurance francaise, 25 rue Châteaudun, 75009 Paris, 1988)
In French; gives legal texts, jurisprudence and changes in the insurance market since the 1978 reform of Civil Code responsibilities and the introduction of *assurance dommage*.

J.L. Meikle and P.M. Hillebrandt, *The French construction industry: A guide for UK professionals*, CIRIA Special Publication 66 (1989)
The construction process. Glossary of French terms.

Le Moniteur, *La sécurité incendie dans les établissements recevant du public*, 3rd edn (Editions du Moniteur, 1988)
An example of the consolidation of official texts in a single publication, in this case fire safety in public assembly buildings.

For French users, CSTB has developed a CD-ROM disc, available by subscription, which carries updated information on French standards and regulations related to construction. It also makes available much of this information through two MINITEL services: MINITEL 36-16 CSTB, and MINITEL 36-17 BATIBASE. CSTB's annual *Catalogue*, as well as the Centre's research reports and studies, lists *Avis techniques* issued, the monographs of NOREX on norms and regulations in a number of foreign countries, European market studies, and the French ministry's *Plan Construction et Architecture* publications. Agence pour la Prévention des Désordres et l'Amélioration de la Qualité de la Construction also publishes a general catalogue.

The Netherlands

Ann Alderson, *The Netherlands: Product approval, quality control and building control procedures*, BRE Report (HMSO, London, 1974)
Earlier study of Dutch approval-and-control organizations, and building control arrangements.

G. Atkinson, *European product certification schemes: Netherlands*, Building Technical File No 26 (July 1989)
Reviews previous arrangements for testing and approval of construction products in the Netherlands; building control and the status of certification; the Dutch Council for Certification (RvC);

certification bodies accredited by RvC; the Foundation for Building Quality (SBK); the Association of Certification Bodies for Building Materials (VECIBIN)

Uniform administrative conditions for the execution of works 1989 (Uniforme Administratieve Voorwaarden voor de Uitvoering van Werken 1989), IBR Netherlands Institute for Construction Law, NL-2508 The Hague; English translation by IBR, The Hague (1990)

VROM, *Technische bouwregelgeving in Nederland en Europa*, VROM Ministerie van Volkshuisvesting (April 1990)
Review in Dutch of recent changes in building law in the Netherlands.

Sweden

Gunnar Essunger and Ulf Thunberg, *Review of the QC Situation in Sweden*, CIB Report 109 (1989)

National Swedish Board of Physical Planning and Building, *General Rules for Type Approval and Production Control* Svensk Byggtjänst, Stockholm (1980); also published a number of Chapters of the Swedish Building Code SBN 1980 in English.
The Swedish Council for Building Research publishes an occasional catalogue of its publications in foreign languages, usually in English.

United Kingdom

Harri Huru (Tampere University of Technology, Finland), *The UK construction industry: a continental view*, CIRIA Special Publication 82 (1992)
A view from Finland of UK construction, including the historical development of the professions and professional institutions; building procurement methods and types of contracts and contract forms.

Other Member States

There are a number of similar publications about construction and construction qiality in other Member States of the European Union, including:

G. Atkinson, *European product certification schemes: Belgium*, Building Technical File No 24 (January 1989)
Organizations involved in product certification, assessment and testing; the Marque-Benor system; the Belgian *agrément*; the UBAC system; certificates of homologation – official confirmation based on Ministerial type-specifications; status of schemes and use of certificated products.

G. Atkinson, *European product certification schemes: Spain*, Building Technical File No 28 (January 1990)
The Spanish construction industry and its market; building legislation and regulation; decennial responsibilities; standards, testing and product certification; technical control; some major organizations involved.

G. Atkinson, *European product certification schemes: Italy*, Building Technical File No 29 (April 1990)
Government organizations; building legislation and regulation; decennial responsibilities; standards, testing and product certification; two major construction industry and standards organizations (ICITE and UNI); some key Italian organizations.

G. Atkinson, *European product certification schemes: Portugal*, Building Technical File No 30 (July 1990)
The Portuguese construction industry; trade in construction materials; two key organizations.

Evelyn Cibula, *Building control in Switzerland*, BRE Current Paper 21/70 (1970)
Legal and administrative background; procedures and enforcement; standards and approvals; technical content of legislation.

J.L. Meikle, *The Italian construction industry: A guide for UK professionals*, CIRIA Special Publication 76 (1990)
Construction professions; contractors and suppliers; planning and building control; the construction process; glossary of Italian terms.

S.D. Reynolds and S. Sheppard, *The Iberian construction industry: A guide for UK professionals*, CIRIA Special Publication 67 (1989)
Part I Spain. Country background; the construction market; the legal and institutional framework; the contracting industry; resources for construction; the building team and professions; the impact of European integration; glossary of Spanish terms. Part II Portugal. The legal and institutional framework; the contracting industry; resources for construction; the building team and professions; the impact of European integration; glossary of Portuguese terms.

Some unresolved issues 12

At the latest by 31 December 1993, the Commission, in consultation with the committee referred to in Article 19 (the EU SCC), shall re-examine the practicability of the procedures laid down in this Directive and, where necessary, submit proposals for appropriate amendment.

89/106/EEC Article 23

Behind the initiatives – national and European – directed at achievement of construction quality, described in this book, are a number of issues yet to be resolved: some inherent in the nature of construction; and some whose origin lies in the Single Market and the European legislation it has brought about, and will bring into being, particularly 89/106/EEC, its Interpretative Documents and ongoing work on preparation of European standards.

During the five years of discussion in the Commission's Standing Committee for Construction and its working groups, in CEN technical committees, in EOTA, and in national groups like the UK Department of the Environment's Joint Advisory Committee (Technical Specifications) and the BSI Panel B/–/2 Conformity and Quality Panel for Construction Products, many issues arising from the implementation of 89/106/EEC have been highlighted. Some arise from the wording of Articles in the Directive, the meaning of which is unclear, or ambiguous; and some because procedures specified are impractical, at least in the short term.

Personnel within the Commission have changed. It takes time to resolve grey areas in responsibilities between the Commission and the European standards-making bodies, and between the Commission and an entirely new organization, EOTA.

Some of the issues listed here are in the process of resolution; others have yet to be resolved. They include:

- CE marking;
- definition of 'least onerous procedure consistent with safety' in conformity attestation, and how far products should be subject to compulsory third-party certification;
- common standards for conformity evaluation;
- definition and responsibility for factory production control;
- elaboration of ER2 in terms of reaction to fire;

- definition of levels and classes in mandates and harmonized European standards and technical approvals;
- omission of an essential requirement for personal security against intruders and threatening behaviour under ER4: *Safety in use*;
- omission of efficiency of artificial lighting under ER6: *Energy economy*;
- roles and responsibilities of EOTA and EOTC, and the status of the latter under 89/106/EEC, if any;
- the different ways in which overlapping directives refer to technical specifications;
- the most effective format for EN standards; and
- achievement of effective enforcement of European legislation in general, and under 89/106/EEC in particular.

12.1 CE MARKING

To help UK manufacturers, in October 1994 the Department of Trade and Industry issued, as an Information Document, a guide to CE marking. It points out that – in simple terms – CE marking denotes that a product conforms to certain EU directives designed to remove technical barriers to trade within the Single European Market. To overcome anomalies between requirements of different directives, a CE Marking Directive has had to be adopted. It has resulted in the amendments to various Regulations that came into force on 1 January 1995. It deals generally with three matters:

- enforcement: where an enforcement authority has reason to believe that a product does not fully comply with a directive, except where the product is unsafe the authority can issue a warning and give the manufacturer the opportunity to take corrective action;
- CE marks: government measures to prohibit misuse of marks; and
- where a product is covered by more than directive, compliance is required for the relevant provisions of all directives.

For construction products, the CE mark must be accompanied by the name or identifying mark of the producer, the last two digits of the year of affixation, the number of the EU certificate of conformity, where appropriate, and indications to identify the characteristics of the product on the basis of the technical specifications, and to confirm that it satisfies all the provisions of 89/106/EEC.

To have a product CE marked may be a commercial advantage to a manufacturer, although sometimes this may be debatable. Some consumers' first experience with the symbol 'CE' was when it became a requirement for the marking of toys put on the market after 1 January 1990. Responsibility for fixing it rested with the manufacturer or, for imported toys, with his agent in the Community. It was supposed to indicate a presumption of conformity with requirements of the 1988 Toy Safety Directive; but, as trading standards authorities have discovered, some CE-marked toys were far from safe, and there was little to prevent the CE mark from being applied negligently, mischievously or fraudulently.

Enforcement depends on how effectively use of the CE mark is policed; and that depends, first, on the kinds of regulations and processes adopted by individual Member States and, second, on the resources available to national enforcement authorities, and their competence. The UK Regulations are elaborately detailed legal documents of little direct interest to practitioners, and suggest that, for some Member States, enforcement could be an opportunity for bureaucratic arrangements that hindered rather than promoted trade.

Furthermore, once specifiers understand that a CE mark is not a quality mark – only that a product is fit for an intended use in terms of safety, health and energy economy requirements in European technical specifications – they will continue to seek from suppliers trustworthy information on the properties and performance of products actually available, supplemented by commercial information on prices, deliveries etc., matters of equal if not of more interest to builders. Well-presented, reliable technical and commercial information, supported by informative marking, will be what specifiers and buyers need. Only in the last resort are they likely to pay notice to a CE mark, and then only when they suspect that a product is not up to standard.

There is nothing to stop a supplier from affixing an informative quality mark on a product, as long as it does not conflict with the 89/106/EEC Article 15.5 requirement that

> Member States shall ensure that the affixing to products or their packaging of marks which are likely to be confused with the CE mark shall be prohibited.

For many years, national certification bodies and industry associations have operated informative quality marking systems such as the 'bar marking' requirement of the UK Certification Authority for Reinforcing Steel. Under the CARES scheme for quality assurance all deformed bars carry a three-part rolling mark that establishes: (a) CARES approval; (b) the country of origin; (c) the mill identity. Bar-markings for CARES-approved steel mills are shown in the list of approved firms. The scheme also distinguishes between the delivery document and the product being delivered, requiring that

> delivery documents for each point of sale will state that the selling organization is CARES approved, and will quote the Certificate Number. The documents will also permit traceability of the material back to its original casts.

As many studies of present certification and marking systems show, the CARES scheme is far from being unique. With the publication of a new edition of the CIB Master List of headings for the arrangement and presentation of information in technical documents for design and construction, in which product performance headings are correlated with the 89/106/EEC Interpretative Documents, now adopted in the British Standard on the presentation of technical information on construction products and services (BS 4940: 1993), there is a well-established system for informative marking of construction products.

12.2 LEAST ONEROUS POSSIBLE PROCEDURE

89/106/EEC Article 13 requires that the procedure chosen for attestation of conformity shall be 'the least onerous possible procedure consistent with safety', while Article 2 requires that construction products 'may be placed on the market only if they are fit for this intended use', the use being defined in a somewhat roundabout way as installation in works subject to regulations containing the essential requirements. Member States tend to give different weight to safety. For example, the German building regulation system places greater weight on structural safety than those of some other Member States. Again, only a few national regulations give sufficient weight to danger from smoke and toxic gases from burning plastics materials in building fires.

For these and other reasons, it is not easy to balance manufacturing economy with attitudes to safety. Regulators and representatives of consumers stress the importance of safety; producers lobby for 'the least onerous possible procedure' for quite legitimate commercial reasons. In between are practitioners, who not only need to strike a balance between risk and economy, but may have a legal as well as financial responsibility for ensuring that the products used do not impair the safety of the works in which they are used. Their need is for reliable technical information on the performance of a product, supported if possible by the technical opinion of an independent authority recognized by their insurers.

12.3 FACTORY PRODUCTION CONTROL

For attestation of conformity to a harmonized European standard prior to E-marking, 89/106/EEC requires what it terms 'factory production control': that is, a system of permanent internal control of production operated by a manufacturer to ensure that his products conform to relevant technical specifications. The system has to be initially inspected and be under continuous surveillance by an approved certification body if the technical specification requires Certification of Conformity, and has to be initially inspected and, possibly, under continuous surveillance by an approved certification body if the technical specification requires Declaration of Conformity by the manufacturer under the first possibility. Under the second and third possibilities the manufacturer is required to operate a system of permanent internal control of production, but it is not subject to initial inspection nor continuous inspection by a third party.

All the elements. requirements and provisions adopted by the manufacturer shall be documented in a systematic manner in the form of written policies and procedures. The documentation shall ensure a common understanding of quality assurance, and enable the achievement of the required product characteristics and the effective operation of the factory production control to be checked.

Guidance Paper 7, a revised version of which was under consideration in November 1994, gives guidelines for 'the performance of the factory production control of construction products'. The guidance is intended to help manufacturers, specification writers, regulators and certification bodies. Although EN ISO 9000 is not a mandatory requirement of 89/106/EEC, compliance with EN ISO 9002 is likely to be deemed to satisfy its factory production control requirements. However, the guidelines given are not intended to be drafted into technical specifications like harmonized European standards supporting 89/106/EEC.

12.4 FITNESS FOR INTENDED USE

Fitness of use for an intended purpose is often impossible to quantify, its assessment being a matter of technical judgement. This judgement may be that of a qualified structural engineer, possibly supported by licensed engineers, as in Federal Germany, or a technical control officer, as in France, based on calculation using specific information of tested properties of materials, guidance in codes like the new series of Eurocodes, and professional experience. In the Commission Communication (Construct 94/118), referred to at the end of this chapter, the long debate as to how to consider 'fitness for use' is discussed against two positions:

- that 'fitness for use' aspects are not directly relevant to the essential requirements, and could be left to national standards; and
- that, although they are relevant, they should not be regarded as a basis for a seventh essential requirement.

In the field of fire safety, there are unresolved issues relating to 'fitness for intended use', such as reliability and serviceability of detection and alarm products and systems, as well as decisions on the 'least onerous procedure', which need clarification either by the Commission, or, better, entrusted to the appropriate CEN technical committee, where the concerns of regulators, practitioners and suppliers may be represented through national standards bodies, supported by research institutions. Or, in the case of new products, decisions are entrusted to EOTA where representatives from bodies like BBA, CSTB and DIBt are experienced to balance fitness and risk.

Incidentally, while it is a responsibility of the Commission to decide which conformity procedures should be included in its mandate to CEN/CENELEC, and where and how they were related to the standard, the decision on which of the methods of attestation should be mandatory to support a CE mark, there is nothing to prevent a supplier for commercial reasons from adopting a more onerous procedure, and having his factory production quality assured by an approved certification body.

12.5 CONFORMITY CERTIFICATION BY BODIES OUTSIDE THE UNION

Member States are only responsible for notified certification and inspection bodies and testing laboratories established within their territories. There are proposals that the Commission should negotiate mutual recognition agreements with third countries on behalf of all Member States. These are at a very early stage, and do not seem to be a priority matter. In the meantime, the only way certification and testing activities can be carried out in a non-Member State is for such activities to be carried out under a subcontract with a notified body within the Union; that notified body, and the Member State that approved it, retain full responsibility for the subcontracted work.

12.6 RECOGNITION OF NATIONAL TECHNICAL SPECIFICATIONS

89/106/EEC Article 4.3 allows a Member State to communicate to the Commission texts of national technical specifications that the Member State regards as complying with the essential requirements. On receipt of the communication the Commission shall forward these national technical specifications to the other Member States, notifying them of those national technical specifications in respect of which there is presumption of conformity with the essential requirements.

A review confined to those British Standards that, from titles and short abstracts in the BSI Catalogue, appear to be related to ER6: *Energy economy and heat retention*, and include methods of test, safety in use, performance characteristics and the like, suggests that there are a number of British Standards that merit, as a whole or in major part, communicating to the Commission as national technical specifications 'in respect of which there is presumption of conformity with the essential requirements'. A review of similar national standards in other Member States, particularly fire safety standards, would probably have a similar result.

However, the Commission has shown little or no interest in using the Article 4.3 route, even as a temporary measure until an appropriate harmonized European standard is available, possibly because of the work falling on Commission officials, and their fear that use of this route might undermine their mandating responsibilities. However, especially in the fields of energy conservation and fire safety (both matters of wide importance within the Union), there are merits in ensuring that selected national standards are given a European dimension under Article 4.3.

A way forward might be for the Commission to entrust a body such as ENBRI (the European Network of Building Research Institutes) with the task of identifying against the related Interpretative Documents which national standards were potential candidates either for recognition under Article 4.3 or as drafts for harmonized European standards.

12.7 89/106/EEC ARTICLE 16: SPECIAL PROCEDURES

In the long transitional stage before 89/106/EEC can be fully implemented, the Commission has given little or no support to the special procedures available under Article 16. A manufacturer wanting to ensure that a product exported to another Member State will be allowed 'free movement and free use of its intended purpose' when there is, as yet, no suitable standard or technical approval recognized at European level, has two choices:

- As is present practice, he can get the product tested and approved for use in that Member State: for example, in Federal Germany by obtaining a *Zulassung* from the Deutsche Institut für Bautechnik, Berlin, or, in France, an *Avis technique* through CSTB.
- If he thinks national procedures too slow or costly, he can fall back on the Special Procedures set out under Article 16.

The course of action then will be as follows:

- First, the manufacturer has to find out what programme of tests and inspections are required by the other Member State (the Member State of Destination).
- He then must apply to his own Member State (producing Member State) – in the UK to the DoE – to initiate Article 16 procedures and designate a body considered competent to undertake the programme.
- DoE will inform the Member State of Destination that it intends to approve this designated body to carry out the required tests and inspections.
- Using the results of these tests and inspections, the manufacturer will be able to claim free movement for his product and free use for its intended purpose in the country of destination.

For the Member State of Destination there are three safeguards:

- It can question the programme of tests and inspections, and if it has misgivings, substantiate them and inform the Commission.

- If it considers the tests and inspections are not being carried out properly, it can tell the producing Member State, which should take corrective action, telling the Member State of Destination what it has done.
- If it still considers the action taken insufficient, it can impose special conditions on the product, or even prohibit its use, telling the exporting Member State and the Commission.

The text of Article 16 suggests that successful use of this procedure depends on cooperation and mutual trust between the two Member States, and the arrangements now being made between DoE and the Spanish industry suggest that, despite lack of enthusiasm by the Commission, it is possible under Article 16 procedures to foster European trade.

12.8 LOCAL PRODUCTS

Building programmes, broadly speaking, fall under one of two heads:

- work for the state, state monopolies, local authorities and near-public agencies; and
- work for private individuals, small and large businesses, charitable organizations etc.

Much of the private sector and, in some countries, a significant part of the public sector work is directed to new and the improvement of one-family housing. Modernization and preservation of historic buildings and sites are also being given much attention in programmes to protect the national and European heritage. While factory-produced products will usually be specified for replacement of electrical services, heating and sanitary equipment, and sometimes, but not always, for repairs to the building fabric, many components like doors, windows and staircases, and materials like roofing tiles, slates, bricks and stonework, will be of local, possibly handicraft manufacture. Again, in rural areas and small towns, many of the products used will be made by handicraft methods to suit local traditions. Under such circumstances, there are and are likely to continue to be significant local or regional variations in techniques and materials. Only some will be placed on a local market. None will qualify for CE marking.

89/106/EEC does not deal directly with local products, but they are the subject of the first of the Standing Committee on Construction's Guidance Papers. These Papers, although not restricted, do not have a wide distribution and are, in the words of their introduction:

- not legal interpretations of the Directive;
- not judicially binding;
- primarily of interest to those involved in giving effect to the Directive; and
- 'may be withdrawn, amended or elaborated at any time'.

Guidance Paper 1 starts by saying that 'local products' must not be prevented from being placed on a national market, and that Member States will apply the 89/106/EEC Article 2.1 requirement that works in which a local product is used meet the essential requirements. However, a local product cannot carry a CE mark, and the notion of 'local' cannot be applied to the whole territory of a Member State – a possible difficulty for Luxembourg. Despite this guidance, 'local products' remain an unresolved issue. Probably it matters little as long as Member State authorities interpret the guidance for practitioners, particularly for artisan builders and those concerned with historic buildings in a common sense manner.

12.9 MINOR AND MARGINAL PRODUCTS

A different, but related issue is the status of minor products. 89/106/EEC Article 4.5 states that the Commission, with the Standing Committee, will 'draw up, manage and revise periodically a list of products which play a minor part with respect to health and safety'. When the list has been published, a manufacturer will be able to 'place on the market' a 'minor product' after issuing a declaration of compliance with the 'acknowledged rule of technology'. Under Article 4.6, such products 'shall not bear the CE mark'.

Early in work on implementing 89/106/EEC, France and Germany reported that their industries disliked such a class of product, while the UK proposed that what was important was a product's end use. After the UK's presentation of a paper in January 1990, it was decided that listing minor parts was not a priority, and attention was turned to a second issue, 'grey area products': that is, products such as a mobile site office, to which the Directive might or might not be applied. Guidance Paper No 3: *Criteria for appreciation of grey area products* was prepared and issued in June 1990. Like the attempt to define and classify minor products, and to deal further with 'local products', the 'grey area' issue has been left largely to Member States to resolve.

For the UK, the DoE, in a special supplement to *Euronews Construction*, has pointed out that, while declaration of compliance with an 'acknowledged rule of technology' is sufficient to justify a minor product being supplied, it is specifically prohibited from bearing the CE mark, and exporters may find its absence a barrier to cross-border trade. Until, therefore, a manufacturer runs into trouble when placing one of his products on the market, the minor product and grey areas issues will remain unresolved, examples of the difficulty of attempting to cover all construction products in a single item of European legislation.

12.10 FIRE SAFETY

A fire protection system differs from other quality attributes, largely because it controls aspects of a building that may never be put to practical test. Yet fire safety is a central requirement in building regulation, as is the fire-testing performance in the assessment and certification of products and building systems. For these reasons, fire protection and the performance of materials in a fire situation have received special attention under 89/106/EEC. Interpretative Document Essential Requirement No 2: *Safety in case of fire*, issued in final form in July 1993, is the longest of the Interpretative Documents, and among the most comprehensive.

Safety in building fires requires regulation of the use and occupation of buildings, their design and construction, including the fire performance of products and installations, and provision for safety of occupants, firefighters and neighbours is demonstrated by the headings in ID 2: *Safety in case of fire*.

Although fire safety was identified soon after a start of work on 89/106/EEC as a critical area for harmonization, the Commission has been slow to take up what may well be the largest obstacle to an open market in construction products. Initially, attention was directed to fire resistance testing. In 1979, after a first less-successful attempt by the Commission to resolve a number of issues, especially that of the reaction to fire of some construction products, an expert group was set up to report on the capabilities and procedures used in fire-testing laboratories in the various Member States. The results of their study were published in 1983 as EUR 8750. However, the report dealt only with the testing of structural building elements, though from the standpoint of harmonized European standards it is the fire performance of construction products placed on the market that is important.

In September 1984, the Commission gave a 15-member study group representing laboratories, fire authorities etc. the ambitious task of reviewing:

- European fire statistics;
- regulations, standards, certification and inspection procedures;
- research; and
- education, training and information.

The group, in its 1986 report, proposed: a directive on safety of buildings, with particular reference to those accessible to the public; and a directive on methods of testing and classifying the fire behaviour and fire resistance of products. Attention to detection, alarm and extinguishing systems was also suggested. It accepted that the testing and classification of the fire behaviour of construction products were not easy to harmonize at a European level, and that some kind of temporary solution was needed to allow free movement of products pending European harmonization of test methods and regulatory requirements.

BOX 12.1   INTERPRETATIVE DOCUMENT ER 2:
SAFETY IN CASE OF FIRE.

Selected list of headings

Fire Safety Strategy

Engineering approach in the field of fire safety

Provisions concerning works:
  loadbearing capacity of the construction
  limitation of generation and spread of fire and smoke within the
    construction works
  prevention of ignition
  limitation of generation and spread of fire and smoke within room
    of origin
  limitation of generation and spread of fire and smoke beyond room
    of origin
  limitation of spread of fire to neighbouring construction works
  evacuation of occupants
  safety of rescue teams

Provisions concerning products:
  Products subject to reaction to fire requirements
  Products for roof subject to fire requirements
  Products within services
  Components of fire detection and alarm installations
  Components of fire suppression installations
  Products and components of smoke control installations
  Products and components of installations for means of escape
  Components for firefighting installations

Performance of products

Attestation of conformity of products

Treatment of working life of construction works

Treatment of working life of construction products

At present, there are important national differences in the character
of fire tests and in the ways that materials and elements of construction
are classified as a result. There are also significant differences in
national building regulations, and in the ways in which fire safety
measures are enforced in new and existing buildings; also in the
specification of classes of building to which these measures apply.
Other matters are differences in national accreditation systems for fire
testing laboratories: some are university institutes; some come directly
under a government department, usually the interior ministry; some,

like the UK's Fire Research Station, are an integral part of a national research establishment; and some are commercial laboratories, possibly associated with national insurance industries. Arrangements for certification, and the status of certificated products and systems, may be directed at meeting the requirements of building and fire prevention regulations, or those of fire insurers concerned with limiting abnormal risks to property.

National test procedures may be based on ISO 834: 1975 *Fire resistance tests. Elements of building construction*; but, possibly because of the time that has elapsed since this international standard was published, the 1975 edition is little more than a skeleton, which Member States have clothed with more detailed, and possibly more restrictive, requirements. In terms of fire safety, these national differences may be of little significance. But the fact that testing procedures, and the ways in which test results are expressed and products and systems of construction are classified, can be widely divergent has important consequences for manufacturers. It means that each Member State has its own particular prescription of fire resistance, and adoption of any one of the present national standards as a unified method of testing could be regarded by other Member States as introducing an unacceptable commercial bias.

In particular, harmonization of reaction-to-fire tests is essential if there is to be an open market in construction products, an open, competitive market only being possible if a product regarded as being 'fire safe' for a specified use in one Member State is acceptable for the same use in all other Member States. Reaction-to-fire tests have always presented difficulties for standards committees. They are essentially attempts to assess burning behaviour, including its contribution to progress of the fire and generation of smoke and toxic gases under standardized and reproducible test conditions, which approximate to one or more stages of a real fire. No fire test, or combination of tests, can guarantee safety in a particular situation. They form only one of many factors that need to be taken into account in assessing the fire performance of a construction and the consequences for the safety of occupants, rescuers and firefighters. Furthermore, it is not only the burning behaviour of construction products that must be considered. Of at least equal importance is that of the contents of a building, including bedding, furniture and furnishings.

As new materials come onto the market, new building situations arise and, in particular, as the lessons of fire disasters are mastered by the authorities, reaction-to-fire tests have to be updated and/or new tests introduced. Consequently, in the UK, France and Federal Germany alone building regulations refer to some 23 different reaction-to-fire tests; a further 14 in Belgium, Italy and the Netherlands make a total of 37 different tests even before the inclusion of fire test references in Denmark, Greece, Luxembourg, Portugal and Spain.

Such a situation suggests that fire testing, and particularly reaction-to-fire tests, is likely to be the Achilles heel in the opening-up of the construction products markets under 89/106/EEC – the more so as national classification systems differ. To resolve the problem requires action by five groups:

- standards bodies;
- bodies responsible for laboratory accreditation;
- certification bodies and national authorities responsible for their accreditation;
- the regulators i.e. building and fire authorities; and
- the fire insurance industry.

One approach, still delayed by funding difficulties, is for the Commission to support a programme of pre-normative research into what has been called a 'robust' solution. A second approach would be for manufacturers to be able to have any national test required by fire authorities and regulators in another Member State carried out in a fire testing laboratory in their home country, with a guarantee of acceptability of results in the other Member State. This would require extensive inter-laboratory cooperation on test and calibration procedures, with each national laboratory being sufficiently well equipped to undertake testing to requirements of other Member States – an achievable though costly solution.

A third approach is through the 'translation of national test results'. A manufacturer would only be required to have performed those tests necessary at his national level and have the results 'translated' by a reference document. It is an approach that has obvious advantages to manufacturers; but, even when suitable translation documents have been agreed, there are a number of disadvantages – commercial and technical – including a level of confidence by regulators unlikely to be achieved in the short term.

The interim solution now proposed is based on the principle of determining the minimum number of national tests that will satisfy the maximum number of regulatory requirements. Now that ID 2: *Safety in case of fire* has been issued, it should be possible to agree an interim test package of largely national tests.

Attention within the Commission has largely been directed at resolving that part of ID 2: *Safety in case of fire* relating to 'reaction to fire'. But fire safety is an issue of wider importance. Now that tourism is a major economic activity within the Union, there is a need to ensure that hotels and other places visited by the public are equally safe in all Member States. Action is necessary, therefore, in the field of European legislation, covering regulations, standards and certification procedures on the fire safety of buildings, with particular reference to buildings accessible to the public. A use of a research facility, like the BRE large 'burn-hall' at Cardington, and the extensive experience of fire research within CIB Working Commission 14, should be directed to an issue that affects Member States in many ways.

12.11 LEGISLATION ON CONSTRUCTION RESPONSIBILITIES AND LIABILITIES

89/106/EEC is directed at construction products, with the aim of removing barriers to their trade, and only indirectly at construction quality, while the procurement directives are only directed at public contracting authorities. Neither is directed towards the responsibilities of constructors to building owners for sound construction, or their liabilities if the works are defective.

An issue debated but as yet unresolved in the Union is the harmonization of the responsibilities of constructors and suppliers to building owners, and their liabilities for damage arising from defective work. The Commission's intention to initiate a construction liability directive has been in the offing for several years.

One proposal has been centred around a specific directive on construction, and the setting-up of a European construction agency either as an extension of the responsibilities of the EC Standing Committee on Construction, set up under 89/106/EEC, or as its replacement by a new standing committee after 31 December 1933, the date by which 89/106/EEC Article 23 requires the Commission, with the Standing Committee on Construction, to 're-examine the practicability of the procedures laid down by the Directive'.

The proposal would have the aim of giving protection to the clients of European construction industries by specifying the responsibilities of constructors and suppliers, and their liabilities for damage occurring during a specific period after completion of works. It would remove many of the uncertainties over present responsibilities and liabilities, which vary significantly between Member States, disadvantage constructors and, to a lesser degree, suppliers, and can be a barrier to trade within the Union. It may be claimed that the harmonization of responsibilities and liabilities would ensure that constructors would be better able to compete for work in the commercial, industrial and private residential sectors on level ground throughout the Union.

In 1987, the Commission engaged a French civil engineer, Claude Mathurin, to carry out an analysis of the ways in which constructors' responsibilities were specified, and their liabilities for building damage insured under national systems. He was later charged with the development of a Community model as the basis for a Construction Industry Specific Directive. M. Mathurin identified a large number of differences in national arrangements caused by:

- systems of national building regulation, their administration and enforcement;
- roles of constructors, particularly the role and responsibilities of architects;
- technical competence of parties involved, and strength and availability of technical services; and

- contractors' liabilities and insurance arrangements, and extent of use of subcontractors.

There was already available much information on the operation of existing national legislation on responsibilities and liabilities through studies that CIB Working Commission W87: Post-construction liabilities and insurance has been undertaking, now collected in a recently published book edited by Jens Knocke, formerly of the Swedish National Institute for Building Research. The French experience is well documented, and there are numerous reports on national regulation systems. There is also more recent experience in operating a building defects fund for social housing in Denmark.

One proposal, on the lines of the Danish scheme, is for a five-year no-fault warranty for all new residential housing. It would involve:

- a latent defects period of five years;
- a start date of the warranty from issue of certificate of practical completion;
- mandatory no-faults warranty on all new residential building;
- warranties to cover material damage and immediate consequential loss;
- joint liability replaced by limiting liability to work for which a party is specifically responsible;
- rights 'inherited' when the building is sold on;
- if the vendor fails to pass on warranty, he remains responsible;
- warrantors retain subrogation rights against constructors, unless waived;
- proof of fault based on reasonable standards of care;
- maximum insurance payment limited to market value of building, after excess; and
- beneficiary person – public housing authority, social housing association, private landlord or owner–occupier – having primary responsibility to repair.

Among issues that need further study are the following:

- Should responsibilities cover breaches of six essential requirements set out in 89/106/EEC, and elaborated in the Interpretative Documents and harmonized European standards?
- Where work is carried out by a subcontractor, should the period of his responsibility start from acceptance of the work by the main contractor, or from the time of overall practical completion?
- Should maximum payment be limited, say to 80% of market value?

A further proposal has been to extend a European warranty scheme to non-residential construction, which might or might not include civil engineering work. Here there are further matters for consideration:

- Should regulators or insurers require certain aspects of the design–construction process to be carried out only by qualified and/or licensed persons or firms?

- Would technical control undertaken to limit insurers' risks be extended to cover essential requirements and/or national regulations, and, if so, should technical controllers be licensed and employed by building owners rather than the insurer, as in France since 1978?
- If so, would technical control be an alternative to, or substitute for regulation by a public authority?
- Should handing over – practical completion or, in France, *réception* – be a formal process with the technical controller carrying a responsibility, either at this stage or at the end of constructors' responsibilities, for discovery for defective work?

The 1991 Mathurin proposals have so far failed to gain strong support from Member States for a Construction Industry Specific Directive. However, a Commission proposal for a Directive on Liability of the Supplier of Services of a general, non-specific character resulted in the appointment of four working groups of GAIPEC (Groupe des Associations Industrielles et Professionnelles), representative of consumer and construction interests, to prepare reports on procedures leading to a single act of acceptance or *réception* of completed work; responsibility and liability of constructors; feasibility of an EU-wide warranty; and the acceptance of such a warranty by insurers.

Up to now, the Commission and Member States have been unable to resolve these matters, or even to make a start by adopting a five-year warranty based on the Danish scheme. It is worthy of note that the creation of a Single Market also involves eliminating barriers in the supply of insurance services, enabling European insurers to offer non-life insurance services in any Member State. With a growing number of insurers participating in the work of the CIB Working Commission W87, it should be possible to agree the framework of a European scheme for damage insurance before the end of the decade, with the foundation built on a five-year warranty or longer period.

12.12 EUROPEAN STANDARDS IN THE VOLUNTARY SECTOR

The status of European standards not mandated by the Commission involves three related issues:

- the relationship between HENs (harmonized mandated standards) in the regulatory sector and ENs (European standards) for voluntary commercial uses;
- in the voluntary sector, the need to strike a balance between the interests of suppliers and users, between large multi-national concerns and small businesses, and between different national interests; and

- the effects of different national testing and certification procedures and interpretation of conformity requirements; also the arbitrary nature of enforcement.

In October 1990, the Commission issued a Green Paper: *The development of European standardization: action for faster technological integration in Europe*. It underlined the importance of European standardization for the internal market on two accounts:

- in technical legislation, where reference to voluntary standards is becoming accepted as the appropriate way of expressing essential requirements; and
- for the economic rationalization needed in an integrated market.

It is becoming increasingly clear that a somewhat arbitrary division in European standards between those parts that are mandated and may be called up in national legislation in which there are essential requirements for safety etc., and those parts that are of commercial interest to suppliers and purchasers but are not directly related to one or more essential requirements, may well be artificial and an obstacle to innovation and economic growth in coming years. The distinction may be of value to regulators but is confusing to many practitioners.

12.13 COMMISSION COMMUNICATION CONSTRUCT 94/118

Some, but by no means all of these outstanding issues have been the subject of Commission Communication Construct 94/118, published by Directorate-General III Industry in October 1994. Under the title 'Standardization and Implementation of the Construction Products Directive', the paper comments, first, on the CPD 'as a tool for the removal of technical barriers to trade'. It identifies national standards as a principal barrier because 'national regulations will refer to these standards when considering essential requirements', and finds that 'most parts of existing national standards are written in descriptive terms'. There follows a section on 'the evolution towards the performance approach', and a further section on 'harmonized standards versus voluntary standards. Two issues are then discussed: 'fitness for intended use', and 'the CPD mandates', followed by 'the answer from the standardization bodies to the mandates'. Three final sections are devoted to: 'the state of the standardization work', 'a procedure for the finalization of the mandates', and 'the CEN working programme for the mandates'. The paper does not discuss the alternative route through EOTA guidelines and European technical approvals.

The distinction made in the Commission paper between a descriptive standard, which is 'what a product is', and a performance standard, which is 'what a product will do', presents difficulties especially with the examples given of coverage of a descriptive standard –

'dimensions and fixings necessary to ensure interchangeability' – and of a performance standard – characteristics of a 'range of products of different materials but intended for a single end use'. Development of a performance approach

> largely depends on the knowledge of the physical, mechanical and chemical phenomena that contribute to the performance of the product.

It goes on to note that when knowledge is sufficiently advanced, as, in a footnote, it may well be in the structural and thermal insulation fields,

> one is able to combine the construction components in different ways, e.g. by calculation, the values that the characteristics of each component should have in order to ensure the performance required for the work.

Such an approach is the route along which modern systems of building regulation are moving. But it requires not only agreement in a standard on which characteristics are important against one or more essential requirements, and how to express and test those characteristics, but also agreement on methods of calculation to assess performance. It therefore means, as in the two fields given as examples, that not only should CEN be mandated to develop harmonized standards covering the characteristics of families of products; there is also a need for harmonized standards covering methods of calculation. They exist in practice in the structural Eurocodes which, however, are not by definition mandated European standards. For other aspects of performance, we have a long way to go before similar Eurocodes are developed.

How, until such Eurocodes are developed, can the performance of families of products, in terms of essential requirements, be assessed is left unanswered. This fact is disturbing, for two reasons. First, British experience suggests that a performance-based regulation system needs two elements: product standards dealing with performance characteristics, and 'design codes' dealing with the calculation of performance for assessment against essential requirements. Second, where the state of knowledge does not yet permit such assessment, an alternative route is available: the *agrément* route.

In the Commission paper there is a second matter that raises issues of importance: that of 'fit for intended use'. While there is agreement that 'fit for intended use' should not be looked on as a seventh essential requirement, there are aspects that are important, such as how far durability affects works or products in service, and the effects of handling and installation operations. Again, in this section, there is reference to 'systems' or 'kits of components put on the market' requiring CE marking, and yet whose declared performance 'applies only if it has been assembled following the assembly instructions of the producers'.

Here, it is useful to refer to the agreement made between the British Standards Institution and the British Board of Agrément in 1983 and, particularly, the matter of products whose performance cannot yet be assessed through a standard procedure but where the materials used are covered in whole or in part by a British Standard. Perhaps, because the Commission paper's main purpose is to review 'the state of the standardization work' in CEN with a view to speeding up the work, it failed to discuss the alternative route to CE marking, that of the European technical approval. An ETA may be granted where 'there is neither a harmonized standard, a recognized national standard, or mandate for a harmonized standard', and where 'a standard could not, or not yet, be elaborated'. Evolution towards the performance approach requires sufficient relevant knowledge to allow performance to be determined by some generally agreed procedure, e.g. by calculation. Until that knowledge is sufficiently advanced to be codified, the ETA route has to be preferred. It means that mandating of ETA guidelines must be given equal weight to that of mandating harmonized standards.

As the quotation from 89/106/EEC heading this chapter states, the Commission, in consultation with the Standing Committee for Construction, was obliged to review 'the practicality of the procedures laid down' in the Directive at the end of 1993. Wisely it was decided that it was too early to make a thorough review, and, instead of a comprehensive review, the Commission devoted resources to completing work in hand. Publication of Construct 94/118 has this as its main objective.

FURTHER READING

Quality marks

George Atkinson, *Quality marks: content and uses*, *CIB 89*, Theme III, Vol. I (June 1989)
Distinguishes quality marks from the limited use of the CE mark.

Building liabilities and insurance

Atkins Planning, *Latent defects in buildings: an analysis of insurance possibilities* (NEDO, London, 1985)

D. Bishop (Chairman), *BUILD: Building Users' Insurance Against Latent Defects*, Report of Insurance Feasibility Steering Committee (NEDO, London, 1988)

John Dobson (Chairman), *Building defects: What can be done and who picks up the bill?* (Legal Studies and Services Ltd, London, 1985)

Jens Knocke (ed.), *Post-construction liability and insurance* (E & FN Spon, London, 1993)
Reviews of current situation in 14 countries.

Claude Mathurin, *Elements for a specific directive on the construction sector* (Paris, 1991)

W.J.R. Smyth (Chairman), *Liability*, Proceedings of the Henderson Colloquium, Cambridge (International Association for Bridge and Structural Engineering (British Group), London, 1984)

O. Zacchi *et al.*, *Quality assurance and liability in building*, Report of a CIB Meeting (National Building Agency, Copenhagen, 1992)

Building regulation

George Atkinson, Some European systems for regulation and control of private design and construction with particular reference to fire safety, *Structural Survey*, **9** (1) (Summer 1990)

DoE, *The future of building control in England and Wales*, Command Paper 8179, HMSO Publications Centre, PO Box 276, London SW8 5DT (1981)
Sets the pattern of the present system, based in principle on maximum self-regulation; minimum government interference; total self-financing; simplicity in operation through the Building Act 1984 and statement of requirements in a series of Approved Documents.

Bengt Eresund, *A survey of building regulations worldwide*, 3rd edn (Byggdok – the Swedish Institute of Building Documentation, Haelsingegata 49, S-113 31 Stockholm,1989)
Entries from 44 countries including Austria, Belgium, Czechoslovakia, Denmark, Federal Germany, Finland, France, Greece, Hungary, Iceland. Italy, Netherlands, Norway, Poland, Portugal, Spain, Sweden, Switzerland, UK and Yugoslavia.

D.C. Mant and J.A. Muir Gray, *Building regulation and health*, BRE Report 97 (1986)

UN Economic Commission for Europe, *Building regulations in ECE countries*, Geneva (1974)
During the 1970s, the UN Economic Commission for Europe, led by Sweden and the USSR, promoted the concept of the unification of national regulations at an international level. Although the study is much out of date, it gave an understanding of differences in aims, content and methods of administration of national legislation and a better understanding of the role played in technical regulations of national standards – mandatory, as deemed-to-satisfy, or just as guides to sound practices.

Fire safety

A list of the extensive UK Fire Research Station publications is given in the *BRE Bookshop Catalogue*.

Conclusion 13

The European Community shares a common inheritance, mostly shared also by these offshore islands: the classical influences on architecture of the Italian Renaissance directly, or through France and the Low Countries; the urban revolution brought about by nineteenth-century railway and industrial development; the use of reinforced concrete as a main construction technique and its codification of its design more or less on a European-wide basis from the 1900s onward; the 'functional' movement in architecture inspired by Le Corbusier and the Bauhaus; the social housing reform movement pioneered by the Netherlands in 1901 and culminating in Weimar Republic Germany (1925–31).

But there are also historical, social and technical differences which explain some of the features of building law and its administration today. First, there is the division between those countries which were in the Roman Empire and those which remained outside, which is reflected in language, law and government – also to some degree in building traditions. There is the later division between Catholic and Protestant Europe, probably a major reason why the traditions and practice of municipal government are stronger in Northern Europe.

More important are the ten years of Napoleonic Empire from 1805 to 1815, which introduced the Napoleonic Code into France and neighbouring countries like Belgium. Napoleon was enamoured by the idea of government as a rationally and scientific constructed system, a matter of technique capable of being applied anywhere regardless of historic tradition. He believed that what people wanted was 'public order, equitable administration, efficient organization' and it could be provided through legal codes and trained administrators.

George Atkinson: 'The roles of authorities, designers and builders in Western Europe', 1970 Chartered Surveyors Annual Conference.

13.1 SHARING A COMMON INHERITANCE

In many ways, since these words were written a quarter of century ago, European constructors, and the civil servants, standards-makers, building scientists, and manufacturers who in their different roles support building and civil engineering, have worked together actively and competitively through the institutions of the Community – renamed since Maastricht the European Union, standards-making bodies like CEN, through a variety of industry and professional organizations, carry through major public works like the Channel Tunnel, and through numerous joint ventures.

There have been disappointments. There has been progress, but not as fast as those who initiated the work would have liked. Economic setbacks; understandable fears, particularly among older citizens recalling past enmities; manoeuvres and jealousies among national politicians; the strength of interest groups: all have contributed to slower than expected growth in cooperation, and in breaking down of barriers.

Working together in European associations, constructors' understanding of the different ways in which works are commissioned, designed and executed in Member States has increased. So has an understanding of the ways in which quality of construction products is regulated, and of how safety and health are protected on construction sites and in completed works. But cultural differences persist. As the introduction to this chapter suggests, they mostly are historical in origin. Those responsible for creating and operating the instruments of cooperation may be criticized for failing to understand such differences. And those whose knowledge of history is based on national accounts of wars and revolutions, and know little of the history of their own institutions, let alone those of other Member States, must also take their share of criticism.

The intent of this book is not to tell the story of European construction and its institutions; but knowledge of how these institutions came into being and how they operate today is necessary. That is why, in Chapter 11, institutions in six Member States are described. The case studies illustrate how Member States regulate building and civil engineering works and control quality of construction products in quite different ways.

The resemblance of European legislation to that of French legislation is suggested. Both share a common base in the traditions of Roman law: understandable when we recall that one of the pioneers of European cooperation was an outstanding French administrator, Jean Monnet, and that another, Jacques Delors, has played a leading role in the transformation from Community to Union after Maastricht. There are disadvantages as well as merits in the Napoleonic concept of a rational system of administration – admirable in its clarity and logic, but liable to degenerate into a dogmatic bureaucracy in the hands of less capable and dedicated men, for whom intervention by public authorities into every aspect of economic and technical activity is an essential feature of public administration. There are also merits and weaknesses in Anglo-Saxon forms of pragmatic management of public business, which, on the one hand, can use the cloak of tradition to justify inaction and, on the other, for reasons of expediency take short-term measures detrimental to construction quality in the longer term.

13.2 TECHNICAL SPECIFICATIONS

The effects of European legislation on construction are described in Chapters 8 and 9. Attention is drawn to an important tool in the creation of the Single Market – the technical specification – a term that, in Commission documents, has acquired a specific meaning, best explained by the following text drawn from 88/182/EEC Annex III:

> Technical specifications: totality of the technical prescriptions contained in particular in the tender documents, defining the characteristics required of a work, material, product or supply, which permits a work, material, product or supply to be described in a manner such that it fulfils the use for which it is intended . . .

Technical prescriptions may include appropriate levels of performance, and, in particular, levels of safety and protection of the health of building users. They may specify requirements applicable to products put on the market or to services supplied, and include a range of matters relevant to their achievement, such as terminology, symbols, testing and test methods, packaging, and marking or labelling. They may include rules relating to design and methods or techniques of construction, and for inspection and acceptance for works. They may also include the training and qualification of practitioners involved in design and construction.

Examples of technical specifications are the many European standards being prepared by CEN/CENELEC technical committees, among which those prepared under Commission mandates – the harmonized European standards – are of key importance. Prepared against the six essential requirements, which reflect the requirements of national regulations and codes, they do not in themselves guarantee construction quality. Even within the important but limited area of health and safety, they need to be supported by design and workmanship codes. As yet, however, these codes, with the exception of the structural Eurocodes, do not exist at a European level but only at national levels, where they may be part of a national system of regulation, or exist as approved guidance documents.

For practitioners and their clients, European technical specifications will grow in importance as they replace national standards. They will, where appropriate, have mandatory use in tender and contract documents for supplies and works under European procurement legislation. And, as they replace national standards, they are likely to have a major influence on construction practice in general, especially if complemented by European design codes.

For practitioners, 89/106/EEC may not be of major importance. Many may find its somewhat cumbersome machinery difficult to understand. However, it should be recalled that its objective is to promote free movement and use of construction products within the Union, and as such it is largely of interest to manufacturers and regulators, and of certification bodies and testing laboratories, which in one way or another are required to attest conformity.

For practitioners, once products meeting European standards or carrying a European technical approval are on the market, the potential value of 89/106/EEC is in promoting 'transparency' in the information that manufacturers provide on the properties of the products that they supply, on their likely performance in specific applications, and where necessary on their proper uses. Here, in parallel, though independent of Union initiatives, publication by CIB of the 1993 edition of the CIB Master List for the arrangement and presentation of technical documents for design and construction is to be welcomed, as is the issue by BSI of BS 4940: 1993 *Technical information on construction products and services.*

Intended to be used in the preparation of many types of document including technical specifications, the 1993 CIB Master List is linked with the process of design, construction, operation, maintenance, repair and supply of construction works, products and services, an important group of headings for information on performance being cross-referenced to headings used in essential requirements and their Interpretative Documents. The revised British Standard provides guidance for those concerned with presentation of technical information on products and services for the UK construction industry, and it is expected that the Standard will be the basis for a wider use as an international or European guide.

Prepared by expert groups from Member States, the six Interpretative Documents have as their immediate purpose to be a bridge between essential requirements set out in 89/106/EEC for safe and healthy buildings, and the drafting of harmonized European standards for construction products. But they could have wider uses, being the foundation of a common 'house style' for European or possibly international construction based on the concept of performance and its achievement through codes and technical specifications aimed at fitness for intended uses, taking into account national traditions, and local social and climatic factors.

The ISO EN 9000 series system standards should also provide the basis of a common European 'house style' for construction industry management, even though it may take time to marry French rationalism, German perfectionism and British pragmatism, not forgetting current practices of other Member States, as well as of three new Member States, Austria, Finland and Sweden.

In creating a European 'house-style' there is the risk that, in the words of a Washington pressman:

> After 1992, Europe will turn in on itself: and become a sprawling, sluggish entity about as relevant to the outside world as the Austro-Hungarian Empire – a decadent pleasure resort.

Some may think this comment from the other side of the Atlantic carries the grains of truth. Even in a sector largely free from immediate political pressures like construction, with the publication of the Interpretative Documents and the more important structural Eurocodes, there could be a degree of complacency in efforts to resolve some of the other issues highlighted in Chapter 12. One area, in particular, stands out for further cooperation: testing for reaction to fire, and the development of a Eurocode for fire engineering.

Possibly because the ISO EN 9000 series of quality system standards, in its earlier form BS 5750, came from a British stable, quality assurance may seem to be an alien, Anglo-Saxon gatecrasher to some German practitioners, used to a comprehensive system of building ordinance controls, and to French constructors with their powerful independent technical control offices. However, there are signs that the European Commission and Member States are taking quality assurance seriously. In October 1992, the Commission issued DOC. Certif. 92/9: *The use of European Quality Assurance Standards (EN 29000 and EN 4500)* as part of the Union's conformity assessment policy. Quality assurance is part of the packet initiated by the Danish National Building Agency for the Building Defects Fund. In France, the central body for public contracts (Commission centrale des marchés) and the council for the Corps des Ponts et Chaussées are supporting quality assurance.

13.3 THE QUALITY OF BUILDINGS

Much of this book is devoted to the quality of construction products, and how European standards are taking the place of national standards in the creating of the Single Market. Less has been said about the extension of the *agrément* concept through European technical approvals, possibly because progress has been slower and has involved bodies whose judgement is generally accepted by all practitioners. That construction products of attested performance make an essential contribution to the quality of the works in which they are used is not challenged. But in itself, quality of construction products is but one of many elements contributing to the quality of buildings and civil engineering works. Unless products are physically and chemically compatible, there will be a loss of efficiency and serviceability, and the life of one or more products used will be shortened, necessitating repair or replacement. Again, the performance of a construction product in a works depends greatly both on how it is stored and handled, and on the skill with which it is incorporated in the works.

However, quality in construction is much more than the selection of compatible products and their use in a workmanlike manner. As early chapters show, achievement of quality starts with the client, whether he is an active participant, or decides to do nothing more than appoint an agent to look after his interest in specifying requirements – explicit or implicit – on his behalf.

Often the client is, in the French, *une personne morale*: a public corporation, board of trustees or limited company. And a lesson drawn from the case studies described in Chapter 3 is that construction quality starts with his ability to choose his professional advisers, and particularly his architect, and to identify and discuss with them his requirements and resources; but quality also depends on the trust that he places on his advisers in aesthetic, economic and technical terms.

Here, for larger projects, there is much to be said for the tradition, exemplified in the Grand Louvre and other major French projects, of distinguishing between the role of the client as *maîtrise d'ouvrage* and that of the project team as *maîtrise d'oeuvre*. The first has responsibilities for determining needs and marshalling financial resources, ensuring that requirements and resources remain in balance as work proceeds and are corrected by changing the mix if they get out of balance; the second has responsibilities that are not only architectural and technical but also economic, in terms of controlling resources and ensuring that any mismatch is corrected.

The 1978 BRE study had started to identify these differences, but their study was never taken further. The manner in which these two linked responsibilities are arranged – sometimes in separate management structures, and sometimes within a single structure – merits further study at a national and European level. So too does the contribution that the quality systems concept could make to their effectiveness. In the 1978 study, it was suggested that to ensure the achievement of quality and value, while all must be done well, preparation of the client's brief – his 'essential requirements' – was the most critical.

FURTHER READING

In July 1994, Sir Michael Latham presented a report, *Constructing the team*, to the UK Department of the Environment. It made 30 recommendations, many of which are related to the achievement of construction quality. They include:

- For clients in the public and private sectors: set up a Construction Clients Forum; select consultants on quality as well as price; publish a construction strategy code of practice.

- For industry: draw up a joint code of practice for selecting sub-contractors.
- For legislation: legislate against unfair contracts; introduce mandatory latent defects insurance.

Organizations, Terms and Definitions

PART 2

International and European Organizations 14

14.1 INTERNATIONAL ORGANIZATIONS

CIB
International Council for Building Research, Studies and Documentation

PO Box 1837, NL-3000 BV Rotterdam, The Netherlands (address: Kruisplein 25g, NL-3014 DB Rotterdam)
Tel: +31 10 411 0240; fax: +31 10 433 4372

Formed in 1953, CIB is the principal international non-governmental organization concerned with technical cooperation in construction matters, mainly through some 90 Working Commissions. With head-quarters in Rotterdam, its members include the world's leading construction research organizations. There are three categories: full members, consisting of the larger national research institutions and industrial R&D organizations; associate members, smaller organizations which either because of the scale of their work or its nature wish to limit participation in CIB activities; and unattached members who are individual specialists.

Most of CIB's work is undertaken in its Working Commissions, results being published as Reports, and, where appropriate, serving as drafts for, or as contributions to international standards. A Memorandum of Understanding has been signed by CIB and ISO concerning cooperation at international level on the use of construction research results in the development of international standards. Working Commissions having a direct interest in construction quality include:

W57: Building documentation and information transfer;
W65: Organization and management of construction;
W87: Post-construction liability and insurance;
W88: Quality assurance;
W96: Architectural management.

Recently W57 issued the 1993 edition of the *CIB Master List of headings for the arrangement of information in technical documents for design and construction*. The first edition was published in 1964 as the *CIB Master List of properties for building materials and products*.

CIB Information, a periodical bulletin, carries news of Working Commissions and Task Groups, CIB publications and other activities.

A new edition of the *CIB Directory of Building Research, Information and Development Organizations* will shortly be published. A list of other CIB publications is available from the Rotterdam headquarters.

ILAC
International Laboratory Accreditation Conference

For information on ILAC, WELAC (Western European Laboratory Accreditation Cooperation) and WECC (Western European Calibration Cooperation), the UK contact is the International Section, NAMAS Executive, 081-943 6554

Established in 1977, the Conference is convened annually. It has the following objectives:

* to promote the exchange and dissemination of information and ideas on laboratory accreditation, laboratory accreditation systems and other arrangements for assessing the quality of test results;
* to facilitate and encourage the acceptance of test results from accredited laboratories, *inter alia* through bilateral and multilateral recognition of laboratory accreditation systems; and
* to cooperate and collaborate with interested international organizations on matters relating to laboratory accreditation and other testing arrangements.

Some 30 countries and 11 international organizations now participate in the work of ILAC, UK interests being coordinated through the NAMAS Executive. The Conference works through a number of Task Forces and Working Groups, one of the most important being that concerned broadly with the development of guidelines on technical aspects of the operation of laboratory accreditation systems, the aim being to develop common criteria against which all national accreditation systems are judged. In this way it is hoped to achieve international recognition of the test results of any nationally accredited laboratory whether for product certification and approval, or otherwise.

Other ILAC Task Forces have been concerned with: criteria for operating internal quality control systems in laboratories; the arrangement of initial and subsequent calibration of laboratory testing equipment; and the operation of proficiency testing programmes.

ILAC publishes periodically a *Directory of Accrediting Systems*. It has been collaborating with ISO in the revision of ISO Guide 25: *Guidelines for assessing the technical competence of laboratories.*

ISO
International Organization for Standardization

1 rue de Varembe, Case Postale 56, CH-1211 Geneva 20, Switzerland
Tel: +41 22 34 12 40

At the international level, ISO and its associated International Electrotechnical Commission (IEC) are recognized by most governments and have as their members the world's national standards bodies. They work closely with regional standards organizations like CEN/CENELEC, and with organizations with common interests, such as ILAC, CIB and RILEM. Similar arrangements exist at a European level between CEN/CENELEC and EOTC, ENBRI and EOTA/UEAtc as well as with national standards, R&D and industry bodies. Recognized by the United Nations and its member governments as the specialized agency for international standardization, ISO has almost all the world's national standards bodies as its participating members, over 90 countries now belonging to the organization. ISO's work is directed to achieving agreement on international standards on a wide range of topics other than in the electrotechnical field, a responsibility of IEC, which shares a headquarters building with ISO in Geneva. Since 1971, the results of ISO standardization work have been issued as international standards, of which there are now over 5000, on topics ranging from earthmoving equipment to information technology. They are listed in an annual *ISO Catalogue*, obtainable either from ISO Publications, Geneva, or from national bodies like BSI. ISO also publishes an *Annual Review*.

International standards serve as a tool for codifying worldwide experience in science and engineering, making the results available to public authorities and industry throughout the world. Of special importance at present is the work on interface standards, which enable links to be established in the structure of information about products used in construction. One example is the lists of 'Agents acting on a building and its parts' and 'Requirements for buildings and building products', based on ISO/DP 6241:1982; another example is the Universal Decimal Classification (UDC) used by libraries throughout the world.

An important step in the harmonization process has been the revision of ISO Guide 2: *General terms and their definitions concerning standardization, certification and and testing laboratory accreditation*. In it the different sorts of bodies responsible for standards and regulations are clarified, and levels and types of standards, stages in the preparation and implementation of standards are defined. Ways of referring to standards in regulations are explained. A section covers testing, certification of conformity and accreditation of testing laboratories, a subject dealt with in detail in ISO Guide 25: *Guidelines for assessing the technical competence of laboratories*.

Following ISO/IEC agreement on criteria for accreditation of certification bodies and laboratories, national accreditation systems such as, in the UK, NACCB and NAMAS are being set up worldwide. A further step towards international recognition was a decision at the 1985 ISO Council and General Assembly to set up a Committee for Conformity Assessment, a topic where there remains a need for better cooperation between ISO and CEN, especially where the CEN work is subject to mandating by the European Union.

RILEM
Réunion Internationale des Laboratoires d'Essais et de Recherches sur les Matériaux et les Constructions (International Union of Testing and Research Laboratories for Materials and Structures)

Domaine de Saint-Paul, F-78470 St Rémy-les-Chevreuse, France
Tel: +33 1 30 85 20 10

Founded in 1947 as a non-profit-making international association, RILEM's principal role is to foster mutual cooperation between member laboratories and standards organizations in the development of test methods, and associated assessment and measurement techniques, relevant to construction. Just under 60 of its 800 members are titular, i.e. representing major laboratories, standards organizations and research bodies, the rest being individuals and firms with a technical interest in RILEM work. From its headquarters in France, RILEM undertakes joint studies aimed at harmonizing test procedures used in construction research, or in the assessment of performance and control of quality of construction products and structures.

In its work RILEM works closely with CIB, periodically organizing joint conferences on topics of mutual interest, such as durability of building materials and components, and appropriate materials for low-cost housing in developing countries. It also cooperates with CEB, the Euro-International Committee for Concrete, on durability and service life of concrete structures, and cooperated with ILAC in the 1990 report: *Test quality for construction materials and structures*, published by Chapman & Hall. ISO and RILEM have a memorandum of understanding, the intention being that RILEM Technical Recommendations will be published as ISO B Documents serving as provisional drafts for development.

UIA
International Union of Architects

51 rue Raynouard, F-75016 Paris, France
Tel: +33 1-45 24 36 88; fax +33 1-45 24 02 78

UN/ECE
United Nations Economic Commission for Europe

Palais des Nations, CH-1211 Geneva 10, Switzerland
Tel: +41 22/34 60 11

UN/ECE is a Regional Commission of the United Nations. Based in Geneva, its membership includes almost every European country. Set up in 1948, the UN/ECE Committee on Housing, Building and Planning has served for over four decades as a forum for the exchange of national, largely governmental experience in urban and regional planning, housing and related building matters. In 1979 it published the study *Building regulations in ECE countries*, in which information on systems of building regulation, procedures for approval of buildings and building products, and 'reference to standards' in technical regulations was brought together from 25 countries. One purpose was to support a UN/ECE Working Party on Building Activity directed at 'international harmonization of approval and control rules for buildings and building products', largely coordinated from Sweden.

The ECE initiative has been overshadowed by developments within the European Union resulting from the 1985 Single European Act, and the decision by Member States in 1989 to adopt the Commission's approach to harmonization under 89/106/EEC, which had the more limited, though still arduous objective, of mandating work in CEN on European standards where EU and EFTA members participate. It is, therefore, doubtful whether there is now any need for the governmental housing and building activities of the UN/ECE Committee on Housing, Building and Planning, at least for European work on harmonization of approval systems for products.

14.2 EUROPEAN ORGANIZATIONS

CEN/CENELEC
The Joint European Standards Organization

rue de Stassart 36, B-1050 Bruxelles, Belgium
Tel: +32 2 519 68 11; fax +32 2 519 68 19

The two European standards organizations, European Committee for Standardization (Comité Européenne de Normalization, Europaische Komitee für Normung – CEN; and European Committee for Electrotechnical Standardization (CENELEC), both autonomous non-profit-making international organizations, share a secretariat in Brussels.

CEN was set up in 1960 and moved from Paris to Brussels in 1985. It has 18 members: the national standards organizations of the EU Member States and EFTA countries. The UK is a member through the British Standards Institution. Membership of CEN is given in Box 14.1.

CEN policy is established through its General Assembly (CEN-AG), on which the heads of national standards bodies, representatives of EU and EFTA governments, and observers from the Commission, EFTA and some other European organizations sit. It is directed by a smaller Administrative Board (CEN-CA), on which only heads of national standards bodies sit. There is a Technical Board (CEN-BT), assisted by three Sectoral Boards – BTS1 (construction); BTS2 (engineering); and BTS3 (health and the environment) – which are responsible for launching and monitoring the work of more than 200 technical committees (CEN-TCs) and supporting working groups (CEN-WGs) manned by delegations from national standards bodies

BOX 14.1 CEN/CENELEC MEMBERS

Austria: Österreichisches Normungsinstitut (ÖN)
(+43 22 22 26 75 35)

Belgium: Institut Belge de Normalisation (IBN)
(+32 3 734 92 05)

Denmark: Dansk Standardiseringsrad (DS)
(+45 3162 32 00)

Finland: Suomen Standardisomislitto r.y. (SFS)
(+358 0 64 56 01)

France: Association Française de Normalisation (AFNOR)
(+33 (1) 42 91 55 55)

Germany: Deutsches Institut für Normung e.V (DIN)
(+49 30 26 01-1)

Greece: Hellenic Organisation for Standardisation (ELOT)
(+30 1 201 5025)

Iceland: Technological Institute of Iceland (STRI)
(+354 1 68 70 00)

Ireland: National Standards Authority of Ireland (NSAI)
(+353 1 37 01 01)

Italy: Ente nationale Italiano di Unificazione (UNI)
(+29 272 00 11 41)

Luxembourg: Inspection des Travaux et des Mines (ITM)
(+352 49 921 21 06)

Netherlands: Netherlands Normalisatie Instituut (NNI)
(+31 15 69 08 90)

Norway: Norges Standardiseringsforbund (NSF)
(+47 2 46 60 94)

Portugal: Instituto Portugués da Qualidade (IPQ)
(+351 1 53 98 91)

Spain: Associcioñ Española de Normalización v Certificación
(AENOR) (+34 1 410 48 51)

Sweden: Standardiserings Kommissionen i Sverige (SIS)
(+46 8 23 04 00)

Switzerland: Schwizerische Normen Vereinigung (SNV)
(+41 1 384 47 47)

UK: British Standards Institution (BSI)
(+44 181 996 7000)

like BSI and drawn from industry, public authorities, and consumer groups. European organizations may apply to join a technical committee or working group but may not vote.

Proposals for work to be undertaken by CEN result either from a mandate from the European Commission, or a proposal from a CEN technical committee, a national member or a European professional or trade organization, for which there is a special procedure. Under the 'standstill agreement' national members must technically stop work on any conflicting standard once agreement is reached to commence work on the topic in a CEN technical committee. For funding, CEN depends in large part on contributions from its members, the European national standards bodies. However, part of its funds comes from the Commission for mandated work under 89/106/EEC.

To take on additional work and otherwise reach agreement on policy and, where necessary, technical matters, CEN operates a system of weighted voting. If ISO is already engaged on a subject, CEN confines its activities to complementing and implementing ISO work. If a draft is accepted by CEN and a European standard is published, the standard is automatically accepted by member countries and must be published as a national standard and the equivalent national document withdrawn, unless the national delegate has abstained and voted against it.

CENELEC is CEN's electrotechnical counterpart, with roughly the same organization – General Assembly, Technical Board and Technical Committees – and similar arrangements with its international

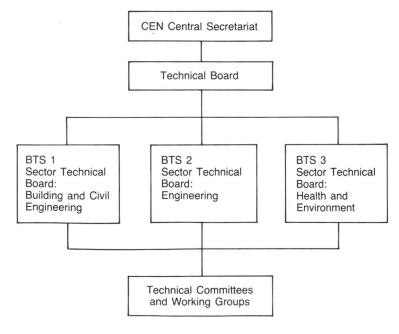

Figure 14.1 CEN organization.

counterpart, the International Electrotechnical Commission (IEC). In the UK, IEC/CENELEC activities are coordinated through BSI's Electrical Department. CENELEC has published a number of electrical safety standards in response to a pre-New Approach low-voltage Directive, which may not meet the needs of New Approach directives. A task yet to be resolved is the preparation of a standardized European electrical plug and socket system.

In matters to do with the implementation of 89/106/EEC and drafting of European standards for construction products, CEPMC (Conseil Européen des producteurs de matériaux de construction – Association of Construction Product Manufacturers in the European Union and EFTA) takes a key interest on behalf of its member bodies: Bundesverband Steine und Erden EC (Germany); Austrian Association for Building Materials; PMS (Belgium); Building Committee of the Federation of Danish Industry; Confederacion Nacional de la Construction (Spain); Finland Cooperative of Building Materials; AIMCC (France); BMP (UK); Building Materials Federation (Ireland); Norske Byggevareprosenters Forening (Norway); Nederlands Verbond Televering Bouw; Industries Byggmaterial Grupp (Sweden).

Box 14.2 shows the three categories of European standards. Other CEN/CENELEC publications include:

- harmonization documents (HDs);
- European pre-standards (ENVs), which may run, like the structural Eurocode series, in parallel with national standards for an agreed trial period; and
- pre-ENs, which are draft European standards issued for comment.

Harmonization documents (HDs) and European standards (ENs), when issued, are listed in *BSI Update*. BSI also lists periodically the appointment of new CEN technical committees, with summary details of the scope of their work, and the national standards body that provides the committee secretariat. Once a harmonized European standard has been agreed, the Commission must publish a reference in the 'C' series of the *Official Journal of the European Committees*.

CISS
European Committee for Iron and Steel Standardization

Until 1986 CISS was responsible for European iron and steel standards. It is now closely linked with CEN, its secretariat being based in the Joint Organization in Brussels.

CEC
Commission of the European Union

rue de la Loi 200, B-1049 Brussels, Belgium
Tel: +33 1-235 11 11

BOX 14.2 EUROPEAN STANDARDS CATEGORIES

Category A

Design installation and execution standards (e.g. structural Eurocodes) Although this category of standards, which equates to codes of practice, plays a key role in the national building legislation of Member States, because the construction product route was adopted to achieve harmonization, preparation of this category of ENs is receiving low priority.

Category B

Product standards of two types:

* HENs (harmonized European standards)
  mandated by the Commission for those characteristics subject to essential requirements listed in 89/106/EEC Annex I
* ENs (non-harmonized European standards)
  standards for characteristics other than those subject to Annex I essential requirements

Whether a HEN or an EN type of product standard, in general, it should be written in performance terms.

Category Bh

'Horizontal standards' for methods of test and measurement which span a number of products

---

The Commission has three responsibilities:

* to ensure that EU rules and principles are respected;
* to propose to the Council measures likely to advance the development of EU policies;
* to implement EU policies

To carry out these responsibilities it has executive powers, some being subject to procedures for collaboration and consultation with national experts. The core of the Union policy-making process is a dialogue between the Commission, which initiates and implements policy, and the Council of Ministers, which takes major policy decisions. Practices throughout the Union are regulated by means of various legislative or quasi-legislative instruments. Matters of concern to construction are examined in Brussels by the Commission, with the help of independent experts, by the Standing Committee for Construction, and by working groups on which Member States are represented, usually by officials from construction and trade ministries (in the case of the UK from DoE and/or DTI).

There are 17 Commissioners, each working through a Cabinet, headed by a Chef de Cabinet, and some 22 Directorates-General, including the General Directorate for the Internal Market and Industrial Affairs, responsible for implementation of 89/106/EEC.

Three other bodies have a special role in the development of the Union and Union legislation.

The EU Council, made up of the Ministers of relevant departments of national governments of the 15 Member States, is the decision-making body on Union legislation. For general matters of policy the foreign ministers of Member States form the Council of Ministers meeting under the Presidency of one of their number, the Presidency rotating among Member States every six months.

The European Parliament at present acts as a consultative and advisory body on Union legislation, but can delay, amend or reject legislation until overruled by the Council of Ministers acting unanimously. The European Court of Justice, based in Luxembourg, is one of four equal institutions referred to in the founding Treaties of the European Communities. In permanent session, the Court's principal task is to ensure that European legislation is applied throughout the Union in accordance with the provisions of the Treaties. Six types of cases come before the Court:

- disputes between Member States;
- disputes between one of the Union institutions and Member States;
- disputes between Union institutions;
- disputes between individuals, or corporate bodies, and Union institutions (including staff cases);
- opinions on international agreements; and
- preliminary rules on cases referred to it by national courts on interpretation of Union legislation, including Directives and Regulations.

Judgments of the Court have consolidated and strengthened Union legislation as much in the interests of individuals, commercial firms and other corporate bodies as that of the governments of Member States or the Commission.

European legislation has started to affect construction in a number of ways. As Box 14.3 shows, it takes on a number of forms.

The route taken in European Community – now European Union[1] – legislation and the respective responsibilities of the different participants are shown in Figure 14.2.

EC SCC
European Commission Standing Committee on Construction

c/o General Directorate Internal Market and Industrial Affairs, 200 rue de la Loi, B-1049 Brussels, Belgium
Tel: +32 2 295 82 56; fax: +32 2 296 10 65

The Standing Committee on Construction, under 89/106/EEC Article 19, is made up of two representatives of each Member State, who 'may

BOX 14.3 FIVE KINDS OF EUROPEAN LEGISLATION

- **Directives**

Most widely known, because of their technical implications; originating in a Commission proposal, adopted by the Council of Ministers in cooperation with the European Parliament, usually after receiving the opinion of the Economic and Social Council.

'A directive shall be binding, as to the result achieved, upon each Member State to which it is directed, but shall leave national authorities choice of corm and methods.'

Article 189(3) of the Treaty of Rome

- **Regulations**

Second type of legal instrument, of general applicability; binding in its entirety and directly applicable to all Member States.

'In order to carry out their task the Council and the Commission shall make regulations . . . A regulation shall have general application. It shall be binding in its entirety and directly applicable to all Member States'

Article 189(3) of the Treaty of Rome

- **Decisions**

The Council or Commission, directly or on the basis of a Regulation or Directive, may address a binding Decision to a government, enterprise or individual.

'In order to carry out their task the Council and the Commission shall (inter alia) take decisions . . . A decision shall be binding in its entirety upon those to whom it is addressed. It may be addressed to Member States or individuals and shall take effect upon such notification. A Member State would, if necessary, rely on national legislation to bring EC Decisions into direct effect'.

Article 189(3) of the Treaty of Rome:

- **Recommendations and Opinions, Case Law**

Recommendations and Opinions may be issued by the Commission or a Directorate-General, the Economic and Social Council or the European Parliament; are not binding but may be useful for the promotion of government or sectoral business at an European level. Case law, result of decisions taken by the European Court of Justice in task of interpretating Directives and other Community legislation

Ref: C: Official Information and Notices, and L: Legislation series of the *Official Journal of the European Communities*

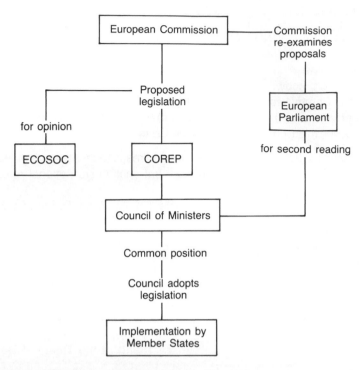

Figure 14.2 The route taken in European Union legislation. ECOSOC: Economic and Social Committee. COREP: Committee of Permanent Representatives of Member States.

be accompanied by experts', chaired by a representative of the Commission. EC SCC met first in March 1989 and, up to July 1994, has met 27 times.

Article 20 gives EC SCC certain tasks and responsibilities:

- to assist the Commission on questions arising from the implementation and practical application of the Directive on Construction Products;
- to assist the Commission in drawing up, managing and revising periodically a list of products that play a minor part with respect to health and safety and in respect of which a declaration of compliance with the 'recognized technical good practice', issued by the manufacturer, will authorize such products to be placed on the market;
- to deliver an urgent opinion when notified either by the Commission or by a Member State that a harmonized standard, a national technical specification submitted by a Member State, or a European technical approval does not satisfy one or more Essential Requirements;
- to advise on instructions to technical committees drafting Interpretative Documents and, when prepared, on their publication in the *Official Journal of the European Communities*;

- to advise on any mandates given by the Commission for establishing harmonized standards and guidelines for European technical approvals;
- to advise the Commission on which of the procedures for applying CE marks for a product or family of products should be specified, taking into account their nature, susceptibility to defects during manufacture, the effects on serviceability of variability in the products' characteristics, and their criticality in relation to essential requirements, particularly in terms of safety and health;
- to advise the Commission that it can grant a European technical approval either because, as yet, there does not exist a harmonized standard, a recognized national standard or a mandate for a harmonized standard, or the product involved differs 'significantly from harmonized or recognized national standards'.

EC SCC is required to help the Commission in the task of monitoring and reviewing European technical specifications 'on a regular basis'; and may be involved where there is a conflict between approval bodies as to whether or not, in the absence of published guidelines, a European technical approval meets the essential requirements. Its advice will also be sought by the Commission when a Member State tells the Commission that a product is being withdrawn from the market because it does not conform with the requirements of the Directive. And, finally, it has the task – 'at the latest by 31 December 1993' – of helping the Commission to re-examine the practicality of the procedures laid down in the Construction Products Directive and, if needed, submitting proposals for any amendment.

Since summer 1992, to speed up the work of the Standing Committee, a smaller Technical Preparatory Group (TPG) has been meeting 4 to 6 weeks ahead of the meeting of the Standing Committee.

EFTA
European Free Trade Association

9–11 rue de Varembé, CH 1211 Geneva 20, Switzerland

Set up in May 1960 as a free trade association of a number of European states not originally signatures to the Treaty of Rome, the Member States of EFTA – Austria, Finland, Iceland, Norway, Sweden and Switzerland – have a population of around 32 million compared with the Union's 322 million. EFTA members share with EU Member States membership of CEN, the European Committee for Standardization, and much of EFTA's activities are now directed at removal of non-tariff barriers within the wider European Economic Area, and maintaining close links with the Union, with which a number of EFTA members, notably Austria, Sweden and Finland have negotiated membership.

ENBRI
European Network of Building Research Institutes

c/o BRE, Garston, Watford WD2 7JR, UK

The European Network of Building Research Institutes comprises nine institutes, and aims to:

- help the Commission on technical aspects of buildings and construction materials, components and systems;
- help firms, and their representative bodies, on the development and suitability of their products and services within the Union or in outside markets;
- give similar help to users and customers of these products and services;
- undertake, through member institutes, research and technical studies needed as back-up to such advice; and
- develop and carry out research and application projects funded with help from the Commission and/or other European bodies.

Its nine members are: CSTC (Centre Scientifique et Technique de la Construction), Brussels; SBI (Statens Byggeforskningsinstitut), Hørsholm, Denmark; IfBt (Institut für Bautechnik), Berlin; CSTB (Centre Scientifique et Technique du Bâtiment), Paris; EOLAS/IIRS (Construction Division, Institute for Industrial Research and Standards), Dublin; TNO-BOUW, Delft, Netherlands; LNEC (Laboratório Nacional de Egenharia Civil), Lisbon; IETcc (Instituto Eduardo Torroja de la Construccion y del Cemento), Madrid; and BRE (UK Building Research Establishment).

EOTA
European Organization for Technical Approvals

Rue du Trône 12, B-1050 Brussels, Belgium
Tel: +32 2 502 69 00; fax +32 2 50238 14

89/106/EEC Annex II.2 requires that:

> The approval bodies designated by the Member States form an organization. In the performance of its duties, this organization is obliged to work in close cooperation with the Commission, which shall consult the committee referred to in Article 19 (i.e. EC SCC) on important matters. Where a Member State has designated more than one approved body, the Member State shall be responsible for coordinating such bodies; it shall also designate the body which shall be the spokesman in the organization.

The organization, EOTA, is based in Brussels, where it has official status under Belgian law. During 1993 a framework agreement between EOTA and the Commission was finalized and common procedural rules were adopted. Arrangements were made for cooperation between EOTA and CEN as equal partners working along the two parallel routes to CE marking specified in 89/106/EEC. EOTA guidelines for products/product families prepared on the basis of Commission mandates would be restricted to products whose fitness was covered by the six Essential Requirements, although manufacturers may seek common European quality labels, for which national ETA bodies could issue technical assessments in non-mandated areas.

EOTC
European Organization for Testing and Certification

rue de Stassart 33, B-1050 Brussels, Belgium
Tel: +32 2 519 68 11; fax +32 2 519 68 17

In July 1989, the Commission issued: *A global approach to certification and testing – quality measures for industrial products* (COM (89) 209). It referred to a new European organization or infrastructure to be set up within the European standards bodies CEN and CENELEC to foster cooperation and mutual recognition within the voluntary area.

EOTC has two tasks: to encourage, foster and manage the development of European certification systems; and to promote in the interests of regulators, suppliers and users mutual recognition of test reports and certificates based on principles and procedures set out in the two key series of European Standards, EN 45000 and EN 29000. It does this through a number of Agreement Groups, where national testing or certification bodies prepare, sign and manage mutual recognition agreements, possibly linked with some form of advisory panel of users and suppliers.

Unlike EOTA, which is concerned with the mandatory sector under 89/106/EEC, and whose coverage is limited to construction products, EOTC's coverage extends to the whole field of economic activity involving testing and certification in the voluntary sector. Input into EOTC is through European national standards bodies, in UK through the BSI Board Committee for Quality Policy. EQS (the European Committee for Quality System Assessment and Certification), set up under a CEN/CENELEC Memorandum of Understanding in December 1989 to overcome the drawback for suppliers of continuing multiple assessment under the EN 29000: *Quality systems* series, has become a specialized committee of EOTC, its ultimate aim being to achieve 'one-stop' conformity assessment, covering testing, inspection, and certification, for Europe.

UEAtc
European Union of Agrément

c/o CSTB, 4 avenue du Recteur Poincaré, F-75782 Paris Cedex 16,
France
Tel: +33 (1) 40 50 24 28; fax +33 (1) 45 25 61 51

Founded in October 1960, UEAtc brings together national organizations active in the field of *agrément*. Not all issue the equivalent of Agrément Certificates themselves, and among those who do the national status of their certificates may differ. The founder members of the Union were from organizations in Austria, Belgium, Federal Germany, France, Italy, the Netherlands, Portugal, Spain and the UK. Later they were joined by SBI (the Danish Building Research Institute) and IAB (the Irish Agrément Board). Recently organizations in Finland, Norway and Sweden joined the Union; and there are now 13 member organizations, all but four belonging to Member States of the European Union. Two institutes in Hungary and Poland are observers.

Qualification for membership is that an organization issues directly, or through another body, *agrément*-type certificates, usually for new construction products. A number of members, such as CSTB, LNEC and SBI, are national construction research organizations. IAB has a separate board but is closely linked with the Irish national standards body. Only BBA (British Board of Agrément) is devoted solely to *agrément* work.

UEAtc's aim is 'to ensure equivalence' of *agréments* issued by its member bodies through the establishment of common rules. Its Guides for Assessment (formerly known as UEAtc Directives) serve as common frameworks for assessing the fitness of new products; and for confirming *agrément* documents issued by one member for use in a second member's country.

1 Following the Maastricht Treaty, the title 'European Union' has replaced that of 'European Communities', and, where appropriate, EU instead of EC. This has led to some difficulties, as many texts, including items of legislation proposed by the EC Commission and adopted by the EC Council of Ministers before 1992, will continue to carry their earlier references, e.g. 89/106/EEC, the Construction Products Directive.

National organizations in Member States of the European Union and European Free Trade Association

15

Data on organizations in 17 Member States of the EU and EFTA concerned with construction, construction quality, standards, certification, testing and similar activities are summarized. With the growth of and changes in European telecommunication services, telephone and fax numbers given here may not always be current, and may have to be checked. With the exception of UK services, national telephone codes are shown as, for example +31 = Netherlands. For communicating from abroad to the UK, the prefix 0 would be replaced by +44.

Spon's European Construction Costs Handbook (First edition, 1992) gives further information on national construction industries, including contractual arrangements and cost data.

15.1 AUSTRIA

A Austria: Republik Österreich   **(EFTA Member)**
Population: 7 681 000 (92/km^2); 57% urban
Land area: 83 855 km^2 Capital city: Vienna (1 531 346)
Standards body: ON Österreiches Normungsinstitut
Tel: +43 222 26 75 35: fax: +43 222 26 75 52
UEAtc member: FGW Forschungsgesellscaft für Materialprüfung
Tel: +43 222 712 62 51: *fax:* +43 222 712 62 51 21

Austrian Embassy: 18 Belgrave Mews, London SW1X 8HV, UK
Tel: 0171 235 3731

Federal State: Bundesländer responsible for building, including building regulation, and much of public works: Burgenland (3.5% national population); Kärnten (7.1%): Nederösterreich (18.8%):

Oberösterreich (17.1%): Salzburg (6.1%): Steiermark (15.5%): Tirol (8.1%): Vorarlberg (4.2%): Wien (19.5%)

National language: German
Currency: Austrian schilling (ASch)
Österreichischer Amstkalendar, published by Verlag an Österreichischen Staatdruckerei, Wien, lists Austrian public bodies

Osterreichisches Normungsinstitut (ONI)
Heinestr. 38, A-1021 Wien 2
Tel: +43 26 75 35; *fax:* +43 26 75 52
National standards body.

Bundesministerium für wirtschaftlichen Angelegenheiten
Stubenring 1, A-101 Wien
Tel: +43 222 7500-0
Federal Ministry of Economic Affairs has federal building responsibilities including FGW, also authorization of testing laboratories and certification bodies.

Forschungsgesellschaft für Wohnen, Bauen und Planen (FGW)
Lowengasse 47, A-1030 Wien 3
Tel: +43 222 712 62 51; *fax:* +43 222-712 62 51 21
FGW, founded in 1956, is a non-profit-making institution whose membership includes the Federal and provincial governments, larger local authorities and accredited testing and research laboratories. It is a full member of the European Union of Agrément (UEAtc).

Verbindungsstelle der Bundesländer
Schenkenstr. 38, A-1021 Wien
Tel: +43 222 66 37 61 0
Liaison centre for Federal States (*Länder*) whose Building Directors' Conference organizes product approvals at State level.

Bau-Dokumentation GmbH
Mittersteig 13, A-1040 Wien
Construction product information centre, part of Heinze GmbH group.

Forschungsgemeinschaft Strassenbautechnik
Argentierstr, 4, A-1021 Wien
Tel: +43 222 505 87 22
Association for highway construction research.

Österreichisches Institut für Bauforschung
An den Langen Lüssen, A-1190 Wien
Tel: +43 222 32 57 88
Non-profit-making building research institute.

Österreichischer verein für Materialprüfung
Karlsplatz 13, 1040 Wien
Tel: +43 222 588 91 01
Association for materials testing.

As Austria is a Federal Republic, building regulation and quality standards remain a State (*Land*) responsibility. The Federal Chancellery, Federal Ministry for Economic Affairs, ONI, FGW and Verbindungsstelle der Bundesländer provide a degree of coordination, as do professional links under the Conference of State Directors.

15.2   BELGIUM

B Belgium: Royaume de Belgique   **(EC Member State)**
Population: 9 927 000 (325/km²); 97% urban
Land area: 30 518 km²
Capital city: Brussels (136 920); Greater Brussels (976 536)
Standards body: IBN/BIN Institut Belge de Normalisation/Belgisch Instituut voor Normalisatie
Tel: +32 2 734 92 05; *fax:* +32 2 733 42 64
ETA body: Direction 'agrément et des spécifications' (DAS) 'Service qualité', Administration pour la réglementation de la circulation et de l'infrastructure, Ministère des communications et de l'infrastructure. Directie Goedkeuring en Voorschriften (DGV), Dienst Kwaliteit, Bestuur voor Verkeer-reglementering en Infrastructuur, Ministerie van Verkeer et Infrastructuur
Tel: +32 2 287 33 33; *fax:* +32 2 287 31 51
UEAtc member: UBAtc/BUtgb Union Belge pour l'Agrément Technique dans la Construction (as above)
CIB full member: CSTC Centre Scientifique et Technique de la Construction
Tel: +32 2 230 62 62; *fax:* +32 2 230 07 29

Belgian Embassy: 103 Eaton Square, London SW1W 4AB, UK
Tel: 0171 235 5422
Kingdom with responsibilities for building, housing and planing delegated to regional governments: Flanders (*Population:* 5 676 194); Wallonia (3 206 165); and Brussels Region

National languages: Flemish (Dutch); French
Currency: Belgian franc (Bfr)

Ministère des Communications et de L'Infrastructure/Ministerie van Verkeer et Infrastructuur (MCI/MVI)
155 rue de la Loi, B-1040 Bruxelles
Tel: +32 2 287 31 54; *fax:* +32 2 287 31 51
National ministry responsible for central government works, and for representation of construction matters in European Community.

Institut Belge de Normalisation/Belgisch Instituut voor Normalisatie (IBN/BIN)
avenue de la Brabançonne 29, B-1040 Bruxelles
Tel: +32 2 734 92 05; *fax:* +32 2 733 42 64

Belgian standards organization; operates Marque-BENOR; conformance to national standards scheme; cooperates in Belgian *agrément* work.

Centre Scientifique et Technique de la Construction (CSTC/WTCB)
rue d'Arlon 53, B-1040 Bruxelles
Tel: +32 2 230 62 62; *fax:* +32 2 230 91 25
Construction research centre of the Belgian National Building Federation, financed by State grant, levy on contractors, and fees; participates in UBAtc *agrément* certification work, as approved testing laboratory; operates an experimental station, Avenue Pierre Holoffe 21, 1342 Limelette; *Tel:* +32 2 653 88 01.

Bureau de Contrôle pour La Securité de la Construction (SECO)
rue d'Arlon 53, B-1040 Bruxelles
Tel: +32 2 230 30 10
Cooperative association established in 1934 to provide technical control services in building and civil engineering; participates in UBAtc certification work.

Union Belge pour l'Agrément Technique dans la Construction (UBAtc/BUtgb), Service de l'Agrément Technique et des Spécifications (ATG/SAS)
Ministère de Communication et de Infrastructure, Residence Palace, 1er étage, rue de la Loi 155, B-1040 Bruxelles
Tel: +32 2 287 31 53; *fax:* +32 2 287 31 51
Closely associated with Régie de Bâtiments, public agency for management of Belgian state buildings; responsible for certification to Régie des Bâtiments type-specification, and, with IBN-BENOR, SECO and CSTC for linked product certification schemes: Marque-BENOR (conformance to national standards) and UBAtc Agrément (assessment of fitness for use). Testing is undertaken either by CSTC or one of the university engineering laboratories.

Centre Belge d'Etude de la Corrosion (CEBELCOR)
avenue Paul Héger, Grille 2, B-1050 Bruxelles
Tel: +32/2-649-63-96
Belgian corrosion centre.

Centre Nationale de Recherches Scientifiques et Techniques pour l'Industrie Cimentière (CRIC)
rue César Franck 46, B-1050 Bruxelles
Tel: 32/2-649-98-50
Belgian cement industry research and technical centre.

Centre de Recherches Routières
boulevard de la Woluwe 42, B-1200 Bruxelles
Tel: +32/2-771-30-80
Belgian highway study and research centre.

Centre Technique de l'Industrie du Bois (CITB)
Chaussée d'Alsemberg 830, B-1180 Bruxelles
Tel: +32/2-377-49-58
Belgian timber industry study and research centre.

Confederation Nationale de la Construction (Nationale Confederatie van het Bouwbedrijf (CNC)
Lombardstraat 34-42, B-1000 Brussels
Tel: +32 2 510 46 11; *fax:* +32 2 513 30 40
Belgian national contractors federation; publisher of weekly journal *Bouw-bedrijf/La Construction.*

Fédération de l'Industrie du Béton Belgique (FeBe)
boulevard A Reyers 207/209, B-1040 Bruxelles
Tel: +32/2-735-80-15
Belgian concrete industry federation.

There is an account of Belgian construction product certification arrangements in *Building Technical File 24* (January 1989).
   CIRA Special Publication 85 (1991), *The Belgian and Luxembourg construction industries. A guide for UK professionals* describes the background, market, legal and institutional framework of Belgian construction industry.

15.3   DENMARK

(DK) Kongeriget Danmark: Denmark   **(EC Member State)**
Population: 5 138 000 (119/km^2; 86% urban)
Land area: 49 093 km^2
Capital city: Copenhagen (466 723)
Standards body: DS Dansk Standardiseringsråd
Tel: +45 39 77 01 01; *fax:* +45 39 77 02 02
ETA body: ETA – Danmark A/S
Tel: +45 42 86 55 33; *fax:* +45 42 86 75 35
UEAtc member: SBI (as above)
CIB full member: SBI (as above)

Royal Danish Embassy: 55 Sloane Street, London SW1X 9SR, UK
Tel: 0171 333 0200

National language: Danish
Currency: Danish Krone (Dkr)
Kongelig Dansk- Hof- og Statskalendar, published by Schultz
Information A/S, Copenhagen, lists government departments and many public institutions

Dansk Standardiseringsråd (DS)
Baunegårdsvej 73, DK-2900 Hellerup
Tel: +45 39 77 01 01; *fax:* +45 39 77 02 02
DS Danish Standards Association, the national standards organization and member of ISO and CEN, issues standards for building products, some referred to in the Danish Building Regulations. All, in principle, may be subject to certification and marking. Certification schemes are voluntary, but for some products are required for building or public safety legislation. DS does not undertake testing, only certification of conformance to a standard.

Boligministeriet
Slotsholmgade 12, DK-1216 Copenhagen
Tel: +45 33 92 61 00
Danish Housing and Construction Ministry.

Bygge- og Boligstyrelsen (NBA)
Boligministeriet, Stormgade 10, DK-1470 Copenhagen K
Tel: +45 33 92 61 00; *fax:* +45 33 92 61 64
National Building and Housing Agency, Ministry of Housing, responsible for social housing, building legislation and construction industry. Approval of building products and equipment is linked to building, gas, water and drainage regulations requiring varying degrees of approvals for materials, equipment or structures. Traditional materials are approved as being in conformity with DS Danish Standards.

Industriministeriet
Slotsholmgade 12, DK-1216 Copenhagen K
Tel: +45 33 92 35 50
Danish trade and industry ministry

Dansk Ingeniørførening (DIF)
Normensekretariatet, Ingeniørhuset, Vester Farimagsgade 29-31, DK-1606 Copenhagen V
Tel: +45 33 15 65 65
Danish Society of Chemical, Civil, Electrical and Mechanical Engineers; since 1893 responsible for most engineering and construction codes of practice, many of which, on public acceptance, are issued as DS (Danish standards) and quoted in the Danish national building regulations. DIF plays a major role in construction quality and testing through Dansk Selskab for Materialprøvning- og Forskning (Danish Association for Material Testing).

Statens Byggeforskningsinstitut (SBI)
PO Box 119, DK-2970 Horsholm
Tel: +45 42 86 55 33; *fax:* +45 42 86 75 35
Danish national building research institute; UEAtc member, and member of CIB and ENBRI; designated Danish technical approval body under the EU CPD. Founded in 1947 under housing ministry, now an independent public institution carrying out work for private and public sectors but still largely government funded through Danish housing ministry. The Danish ETA body, ETA-Danmark A/S (PO Box 54, DK-2970 Horsholm) is associated with SBI.

Statens Tekniske Prøvenaevn (STP)
Bregåde 31, DK-2200 Copenhagen K
Danish National Testing Board, set up by the Ministry of Commerce in 1973, accredits testing laboratories, its role being similar to that of NATLAS in the UK. Since 1982 STP has maintained a register of accredited laboratories, a condition of accreditation being that the laboratory will insure to cover client losses that might result from faulty testing. The register gives information in English on accreditation schemes. For building and construction, STP has an industry committee: Authorisationudvalget for byggeindustri.

Asfaltindustriens Vejforskningslaboratorium
Stamholmen 91, 2650 Hvidovre
Tel: +45 31 78 08 22

Danish Institute of Fire Technology
Datavej 48, DK-3460 Birkeroed
Tel: +45 45 82 00 99; *fax:* +45 45 45 82 24 99
Joint organization of SKAFOR (Danish Non-life Insurance Underwriters), Dantest (National Institute for Testing and Verification), and the Danish Fire Protection Association, as an independent non-profit making body.

Dansk Institut for Prøvning (Dantest)
Amager Boulevard 115, DK-230 København S
Tel: +45 31 54 08 30; *fax:* +45 31 95 47 00
also Smedegade 14, DK-6000 Kolding
Tel: +45 5 52 37 27;
Østre Gjesibgvej 5-7, DK-6715 Esbjerg (*Tel:* +45 5 13 59 00)
Dantest, the National Institute for Testing and Verification, is an independent non-profit-making test house and metrology institute founded in 1896. Dantest is an STP-accredited laboratory supported by public grants and fees for testing and consultancy services. Its building materials division provides the secretariat for Betonelementkontrollen (concrete components quality control), and Lettbetonkontrollen (light-weight concrete quality control). Fire testing work is now a responsibility of the recently established Danish Institute of Fire Technology.

Dansk Beton Industrieforening (DBI)
Nörre Voldgade 106, DK-1015 Copenhagen K
Tel: +45 1 33 13 88 0; *fax:* +45 1 33 13 24 50

Danmarks Electriske Materialkontrol (DEMKO)
PO Box 514, Lyskaer 8, DK-2730 Herslev
Tel: +45 42 94 72 66; *fax:* +45 42 94 72 61
Danish Board for Approval of Electrical Equipment is a public institution whose certificates are required under electricity regulations. DEMKO undertakes testing, conformity certification and inspection of quality control procedures in the manufacturing process. Approved equipment is D-marked.

Danmarks Gasmateriel Prøvning (DGP)
Tranegårdsvej 20, DK-2900 Hellerup
Tel: +45 31 62 43 85
Danish Gas Equipment Testing Organization, an STP-accredited laboratory certifies gas equipment for conformance to requirements of the 1981 Gas Regulations. Approved gas equipment is DG-marked.

Dansk Isolerings Kontrol
Nørre Voldgade 34, DK-1358 København K
Tel: +45 1 15 17 00

Kalk- og Tegelværkslaboratoriet (K&T)
Teglbaeksvej 20, DK-8361 Hasselager
Tel: +45 6 28 38 11
STP-accredited laboratory of the brick, tile and lime industries, which manages
Dansk Mørtelkontrol, a voluntary scheme controlling the quality of mortar
used in masonry work, which follows DIF procedures for control schemes,
and Dansk Murstenskontrol, a similar scheme for clay and sand lime bricks.

Danmarks tekniske Højskole (DtH)

Laboratoriet for Bygningsmaterialer (LfB)

Afdelingen for Bærende Konstruktioner (ABK)

Laboratoriet for Varmeisolering (LfV)
DK 2800 Lyngby
Tel: +45 32 88 35 11

**Dansk Vindueskontrol: Kontrolordning for vinduer og yderdør
(Danish Control Secretariat for Windows and External Doors)**
Sekretariatet for FDV/KVY Byggeteknik, Jysk Teknologisk, Tekno-
logiparken, DK-8000 Ärhus
Tel: +45 6 14 24 00

Dansk Entreprenoer Foreningen
Nörre Voldgade 106, Copenhagen 1358
Tel: +45 31 13 88 01
Danish Contractors' Association.

Godkendelsessekratariatet for materieler og konstruktioner
Postboks 54, DK-2970 Hørsholm
Tel: +45 42 86 55 33
Approval secretariat for construction products for NBA.

Godkendelsessekratariatet for vand- og afløbsmateriel
Postboks 54, DK-2970 Hørsholm
Tel: +45 42 86 55 33
Approval secretariat for plumbing and sanitaryware for NBA.

**Lamineringsudvalget sekretaeren, Afdelingen for baerende
konstruktioner**
DTH Bygning 118, DK-2800 Lyngby
Tel: +45 42 88 35 11

Lydteknisk Institut
Building 356, Akademivej, DK-2800 Lyngby
Tel: +45/2 93 12 11
Danish Acoustical Institute, Academy of Technical Sciences, accredited by
STP to undertake laboratory tests of sound insulation index of windows and
doors, walls and floors; impact sound insulation for floors; absorption coeffi-
cient for sound-absorbing materials against general and specific requirements
of Danish building regulations.

Nordisk Forskningsinstitut for Maling og Trykfarverågern
Allé 3, DK-2970 Hørsholm
Tel: +45 42 57 03 55
An STP-accredited laboratory for coatings, paints, polymers and adhesives.

Prøvnings udvalget for olietanke Korrosionscentralen
Parkallé 345, DK-2605 Brøndby
Tel: +45 42 63 11 00
Korrosionscentralen is an STP-accredited laboratory operating a scheme for approval of oil storage tanks to prevent ground pollution through leakage.

Teknologisk Institut
Afdelingen for Byggteknik, Gregersenvej, PO Box 141, DK-2630 Tåstrup
Tel: +45 42 99 66 11; *fax:* +45 42 99 54 36
The Technological Institute, one of three principal laboratories accredited by STP, manages Danish industry product quality control schemes, as well as providing testing and consultancy services. Its Department of Building Technology provides the secretariat for a number of product quality schemes: Dansk Fingerskarringskontrol (structural timber finger joints); Fabriksbeton-kontrollen (ready-mixed concrete); Gulvbranchens Vådrumskontrol (water-proof flooring); Trærådets Brandimpraegnersudvalg (fire retardant treatment); Træpladekontrollen (timber-based boards); Træspærkontrollen (timber roof trusses); T-virkeordningen (structural timber grading); Varmeisolerngskon-trollen (thermal insulation materials); Dansk Impraegnerskontrol (timber preservation).

The Institute also shares the secretariats of concrete factory quality schemes with Dantest, and window production control with the Jutland Technological Institute.

There is an account of construction product certification arrangements in *Building Technical File 25* (April 1989); and in a 1974 BRE report by A. Aldersonon *Danish Product approval, quality control and building control.*

15.4   FEDERAL GERMANY

Post codes are in the process of change and may have to be confirmed with the individual organization.

**D Federal Germany: Bundes Republik
Deutschland**   **(EC Member State)**
Population: 79 220 000 (222/km^2); 86% urban
Land area: 108 333 km^2
Capital city: Berlin: 3 115 473
Standards body: DIN Deutsches Institut für Normung eV
Tel: +49 30 26 01-1; *fax:* +49 30 26 01-2 31
ETA body: DIBt Deutsches Institut für Bautechnik
Tel: +49-30 26 48 70; *fax:* +49-30 26 48 73 20
UEAtc member: Din (as above)
CIB full members: DIBt (as above), IRB

Federal German Embassy: 23 Belgrave Square, London SW1X 8PZ, UK
Tel: 0171 235 5033

National language: German
Currency: Deutschmark (DM)

Postal codes: Following unification, German postal codes have been rearranged. UK residents can obtain in English free of charge information on any changed code by dialling 0800 960074. There are also available, in English, copies of the new code book, *Das Postleitzahlenbuch*, free of charge, from Post Dienst Service Centre, Aennehenstr. 19, D-53177 Bonn (*fax:* 01049 228 817 366). There have also been changes in some phone numbers, especially in the new *Länder*.

For current information on addresses, phone and fax numbers, etc, two annuals, in German, may be consulted: *Verbande Behörden Organisationen der Wirtschaft*, published by Verlag Hoppenstedt Gmbh, Postfach 100139, D-64201 Darmstadt; and *Taschenbuch des öffentlichen lebens Deutschland*, published by Festland Verlag, Bonn.

The first lists, under *Technische überwachungsvereine, Materialprüfungsämter, Vermessungsverwaltungen*, the 48 members and associates of VMPA Verband de Materialprüfungsämter and other bodies offering testing and quality control services. It also lists professional organizations of architects and engineers, and organizations concerned with housing, town planning, and building and building materials industries.

Deutsches Institut für Normung eV (DIN)
(Federal German standards organization)

PO Box 1107, Burggrafen Str 4-7, D-10787 Berlin
Tel: +49-30 26 01-1; *fax:* +49 30 26 01-2 31
Member of ISO and CEN. Component organizations include: Deutscher Normenausschuss (DNA/DIN – board of DIN; and Normenausschuss Bauwesen im DIN Deutsches Institut für Normung eV (NABau im DIN), responsible for DIN construction standards including ETB series.

From October 1990 DIN assumed responsibility for standardization throughout Federal Germany, representing unified Germany within CEN, DIN publishes a two-volume catalogue of German standards, most entries being also in English.

Deutsches Institut für Gütesicherung und Kennzeichnung eV (RAL)

Sieberger Strasse 39, D-53757 Sankt Augustin
Tel: +49 2241 16 05 0; *fax:* +49 2241 16 05 11
Independent commission of DIN, financed jointly by industry and Bundesministerium für Wirtshaft; regulates use of quality marks.

Bundesanstalt für Materialforschung und -prüfung (BAM)

Unter den Eichen 87, D-12205 Berlin
Tel: +49-30 81 04 - 0; *fax:* +49 30 811 - 20 29

Federal materials research and testing institution covering all industrial materials.

Deutsches Institut für Bautechnik (DIBt)
Reichpietschufer 74-76, D-10785 Berlin
Tel: +49-30 264 87-0; *fax:* +49-30 264 87-3 20
Founded in 1968 jointly by the Federal and Lander Governments under special legislation, and given tasks: (a) issue of technical (*zulassungen*) on behalf of *Länder* building ministries for new materials, components and systems not covered by published standards against *Länder* building laws and regulations; (b) official recognition to industry laboratories and quality associations set up to control manufactured building products; (c) participation in development of technical building regulations and standards as national, European and international levels, DIBt is the body authorized to issue European technical approvals in Federal Germany.

Bundesministerium für Raumordnung, Bauwesen und Stadtebau
31-37 Deichmanns Av, D-53179 Bonn
Tel: +49 228 337 00; *fax:* +49 228 337 30 60
Federal ministry with overall responsibilities for physical and urban planning, and construction; provides one of two German members of the EU Standing Committee for Construction; has responsibilities for AGB Geschaftstelle der Arbeitsgemeinschaft für Bauforschung – working group for building research

Bundesministerium für Verkehr
Robert Schumann Platz 1, D-53175 Bonn
Tel: +49 228 300 - 0; *fax:* +49 228 300 34 28
Federal ministry for transport, including motorways; provides one of two German members of the EU Standing Committee for Construction.

Each of the 16 States (*Länder*) in the unified Federal Germany (*Bund*) has responsibilities for construction, building regulation and testing and approval of construction products, usually in association with a material testing laboratory attached to a local technical university or one of its research institutes. The German standards body (DIN) has responsibility for standards throughout Germany. *Land* building ordinances are based on a common model, *Musterbauordnung*, BAM, the Federal materials research and testing establishment, DIBt, the Institute for Building Technology, and IRB, the Stuttgart centre for regional planning and construction, also serve the unified Germany. In the following sections, the name of the *Land* government department responsible for construction is given, or, alternatively, the interior ministry (*innenministerium*); also principal research institutes and testing laboratories.

Land Baden-Wurttemburg

Population: 9.2 million; land area 35 571 km²
Seat of **Land** *government:* Stuttgart
Population: 559 000

Finanzministerium, Baden-Wurttemberg, Neues Schloss 4, D-70173 Stuttgart (*Tel:* +49-711 279 0; *fax:* +49-711 279 83 93)

Innenministerium, Dorotheenstr, 6, Stuttgart (*Tel:* +49-711 20 72-1; *fax:* +49-711 20 72 37 29)

Staatliches Hochbauamt 1, Ulrichstr, 3, D-70173 Stuttgart (*Tel:* +49-711 20 72-1; *fax:* +49-711 20 72-37 19)

Technische Universitat, Keplerstrasse 7, Stuttgart 1

Institut für Baukonstruktionen (*Tel:* +49-711 20 73 1)

A number of TU institutes are at Vaihingen, D-70569 Stuttgart: Inst. für Baustatik, Pfaffenwaldring 7; Inst. für Grundbau und Bodenmechanik, Pfaffenwaldring 35; Inst. für Baubetrieblehre, Pfaffenwaldring 7; Inst. für Kunststoffprüfung Pfaffenwaldring 32 (*Tel:* +49-711 685 2-627); Inst. für Werkstoffe im Bauwesen Pfaffenwalrding 4 (*Tel:* +49-711 685 3-324); MPA Staatliche Materialprüfungs-animal, Universitat Stuttgart, Pfaffenwaldring 32 (*Tel:* +49-711 784 1); Amtliche Forschungs- und Materialprüfrungsanstalt für Bauwesen (Otto-Graf-Institut), Pfaffenwaldring 4 (*Tel:* +49-711 685 3323; *fax:* +49-711 685 6820)

Also at Vaihingen:

IBP Fraunhofer–Institut. für Bauphysik (Fraunhofer Institute for Building Physics) (*Tel:* +49-711 686 8-320); IRB Informationszentrum RAUM und BAU des Fraunhofer-Gesellschaft (Fraunhofer Information Centre for Regional Planning and Building) Nobelstr. 12 (*Tel:* +49-711 686 8-500)

Other centre

Karlsruhe
Population: 267 000

Technische Universitat:
Lehrstuhl für Ingenieurholzbau und Baukonstruktionen, Schall-technisches Laboratorium; Lehrstuhl für Stahl- und Leicht-metallbau; Versuchanstalt für Stahl, Holz und Stein, Amtliche Material-prüfungsanstalt, Postfach 6980, Kaiserstrasse 12, D-76131 Karlsruhe (*Tel:* +49-72 608 22 11; *fax:* +49-72 607 3 69) Inst. für Beton und Stahlbeton, Postfach 6380, Am Fasanengarten, (*Tel:* +49-72 608 22 77)

Freistaat Bayern (Bavaria)

Population: 11.2 million; area 70 554 km^2
Seat of **Land** *government:* München
Population: 1 300 000
Innenministerium Odeonsplatz 3, D-80339 München (*Tel:* +49-89 21 92 01; *fax:* +49-89 28 20 90)

Technische Universität: Forschungsinstitut für Bituminöse Baustoffe (*Tel:* +49-89 811 21 05 1); Inst. für Stahlbau, (*Tel:* +49-89 811 21 05 445/6) Materialprüfungsamt für das Bauwesen der Technischen Universität Archisstrasse 21, D-80333 München (*Tel:* +49-89 21 05 30, 39; *fax:* +49-89 21 05 30, 57) Bauzentrum München der Münchener Messe- und Ausstellungs GmbH (Munich Building Centre), Messegelande-Sud, München 70 (*Tel:* +49-89 510 7-441/442; *fax:* +49-89 5107-166)

FhG Fraunhofer-Gesellschaft: Zentralverwaltung (Headquarters of Fraunhofer Foundation), Leonardstr. 54, München 19 (*Tel:* +49-89 120 5-01)

IFO–Institut für Wirtschaftsforschung (IFO–Institute for Economic Research), Poschinger Str. 5, München 86 (*Tel:* +49-89 92 241)

Bayerische Staatliche Prüfamt für Technische Physik, James-Frank-St, Garching bei München (*Tel:* +49-89 3209 24 52)

Forschungsinstitut für Wärmeschutz eV (Research Association for Energy Conservation), Grafelfing, München

Other centres:

Nürnberg
Population: 478 000

LGA Landes-gewerbeanstalt Bayern, Postfach 3022, Gewerbe-museumsplatz 2, Nürnberg 1 (*Tel:* +49-911 201 71)

Rosenheim Bayern
Population: 54 000

Institut für Fenstertechnik eV (Institute for Window Technology), Arnulfstrasse 13, Rosenheim-Aisingerwies (*Tel:* +49-911 65 245)

Land Berlin (incorporating East and West Berlin)

Seat of **Land** *government:* Berlin
Population: 3.4 million; area 884 km^2

Senats verwaltung für Bau- und Wohnungwesen, Wurrtembergstr. 6, D-10707 Berlin (*Tel:* +49-30 8 67-1; *fax:* +49-30 67-73 31)

BMBau Berlin (Berlin Office of Federal Building Ministry), Scharenstr. 2-3, D-10178 Berlin

Unweltbundesamt (Federal Office for Environmental Protection), Bismarckplatz 1, Berlin (*Tel:* +49-30 89 03-1)

BAM Bundesanstalt für Materialforschung und -prüfung (Federal Materials Testing Establishment)

DvM Deutsche Verband für Materialprüfung eV

DIN Deutsches Institut für Normung eV

DIBt Institut für Bautechnik

DITR Deutsches Informationszentrum für technische Regeln

(see under Federal German organizations)

Technische Universität:
Institut für Baukonstruktionen und Festigkeit (Institute for Structural Design and Strength of Materials) (*Tel:* +49-30 314 2980); Inst. für Krankenhausbau (Institute for Hospital Building), Strasse des 17 Juni 135, Berlin (*Tel:* +49-30 314 29 60)

Inst. für Technische Akustik (Institute for Acoustics), Jebenstrasse 1, Berlin (*Tel:* +49-30 314 29 31/32)
Lehrstuhl und Inst. für Städtebau (Town Planning Institute), Hardenbergstrasse 35, Berlin (*Tel:* +49-30 310 78 1); Inst. für Lichttechnik (Institute for Lighting), Einsteinufer 19, Berlin (*Tel:* +49-30 314 24 01); Inst. für Heizung und Luftung (Institute for Heating and Ventilation), Marchstrasse 4, Berlin-Charlottenburg

THA Treuhandanstalt, Leipzgerstr. 5-7, Berlin

Land Brandenburg

Population: 2.7 million; area: 29 059 km^2
Seat of* Land *government: Potsdam
Population: 140 000
Ministerium für Staatenwicklungen, Wohnen und Verkehr, Dortustrasse 30-33, D-14467 Potsdam (*Tel:* +49-331 18 66 10; *fax:* +49-311 2 41 81)

Freie Hansestadt Bremen

Population: 0.6 million; area: 404 km^2
Seat of* Land *government: Bremen
Population: 530 000
Senator für das Bauwesen, Postfach 107647, Ansgaritorstrasse 2, D-28195 Bremen (*Tel:* +49-421 3 61 0; *fax:* +49-421 3 61 20 50)
LAB Landesamt für Baustoffprüfung, Paul-Feller-Strasse 1, D-28199 Bremen

Freie und Hansestadt Hamburg

Population: 1.6 million; area 754 km^2
Seat of* Land *government: Hamburg
Baubehörde, Postfach 30 05 80, Stadhausbrücke 8, D-20355 Hamburg (*Tel:* +49-40 349 13 1; *fax:* +49-40 349 13 7 35)
Bundesforschungsanstalt für Forst- und Holzwirtschaft (Federal Research Establishment for Forestry and Timber Products), Leuschnerstrasse 91, Hamburg (*Tel:* +49-40 739 19 1)
Strom- und Hafenbau Baustofftechnik, Veddeler Damm 16, Hamburg
Bundesvereinigung für Prüfingeneure für Baustatik, Jungfernstieg 36, Hamburg (*Tel:* +49-40 350 09 0; *fax:* +49-40 350 09 1 00)

Land Hesse

Population: 1.6 million; area: 23 114 km^2
Seat of *Land* government: Wiesbaden
Population: 260 000
Ministerium für Landesentwicklung, Wohnen, Hölderlinstrasse 1-3, D-65187 Wiesbaden (*Tel:* +49-611 8 17-1; *fax:* +89-611 84 16 49)

Other large centres

Darmstadt
Population: 135 000
Technische Universität, Karolinenplatz 5, Darmstadt Fachgebiet Strassenbau, Peterstr. 30 (*Tel:* +49-6151 16 25 46)
Lehrstuhl und Institut für Massivbau, Alexanderstr. 5
Lehrstuhl und Institut für Statik und Stahlbau, Alexanderstr. 7 (*Tel:* +49-6151 16 21 45)
Institut Wohnen und Umwelt GmbH (IWU) Institute for Housing and Environmental Protection), Annastr. 15 (*Tel:* +49-6151 26 911)

Frankfurt-am-Main
Population: 617 600
Battelle-Institut eV: Postfach 90 01 60, Am Römerhof 35, Frankfurt (*Tel:* +49-69 790 80)
Oberprüfungsamt für die höhrer Technische Verwaltungbeamte (Office for Examinations to the Official Higher Technical Service), Bockenheimer Anlage 13, Frankfurt-am-Main

Kassel
Population: 189 000
Amtliche Prüfstelle für Beton, Universität Gesamthochschule Kassel, Mönchebergstr. 7 (*Tel:* +49-561 8 04 26 01)

Land Mecklenburg-Vorpommern (Mecklenburg-West Pomerania)

Population: 1.96 million; area: 23 838 km^2
Seat of **Land** *government:* Schwerin
Population: 130 000
Innenministerium, Arsenaal am Pfaffenteich, D-19048 (*Tel:* +49 385 5 88-0; *fax:* +49 385 5 81 1180)

Land Niedersachsen (Lower Saxony)

Population: 7.2 million Area: 47 438 km^2
Seat of **Land** *government:* Hannover
Population: 500 000

Land Niedersachsen Innenministerium, Lavesalles 6, D-30169 Hannover (*Tel:* +49-511 12 0-1; *fax:* +49-511 1 189 962)

Technische Universität:
Institut für Baubetrieb und Baubetriebwirtscaft (Institute for Building Management and Economics), Callinstrasse 32; Institut für Statik, Callinstrasse 15; Institut für Baumechanik; Institut für Baustoffkünde und Materialprüfung; Institut für Electrowärme, Am Welfengarten 1; Institut für Materialprüfung und Forschung des Bauwesen, Nienburger Strasse 3 Institut für Raum und Bauakustik, Schlosswender Strasse 1

IfB Institut für Bauforschung EV (Hannover Building Research Institute), An der Markuskirche 1
Institut für Bauschadensforschung (Institute for Research into Building Failures), c/o Vereingte Haftpflichtversicherung V.a.G: Constantinstrasse 40

Other centres

Braunschweig Niedersachsen
Population: 252 000

Technische Universität:
Carolo-Wilhemina, Pockelstrasse 14 (*Tel:* +49 531 391 4111)

Institut für Baustoffe, Massivbau und Brandschutz (Institute for Building Materials, Reinforced Concrete and Fire Protection), Institut für Baustoffen und Stahlbetonbau; Institut für Stahlbau (Institute for Structural Steelwork), Beethovenstrasse 51-52
Materialprüfanstalt das Bauwesen, Beethovenstrasse 52
Abteilung Kunststoffe, Massivbau und Brandschutz, Hopfengarten 20
Institut für Grundbau und Bodenmechanik (Institute for Foundation Engineering and Geotechnics), Gausstrasse 2
Institut für Industriebau; Lehrstuhl und Institut für Städtebau, Wohnungwesen und Landesplanung, Pockelsstrasse 4

Celle Niedersachsen
Population: 71 500
Heinze Gmbh (Deutsche Bau-Dokumentation), Postfach 505, Bremerweg 184 (*Tel:* 49-5141 500; *fax:* 49-5141 501 04)

Clausthal-Zellerfeld Niedersachsen
Population: 17 100
Amtliche Materialprüfanstalt für Steine und Erden, Zehntnerstr 2A

Nordrhein-Westfalen (North Rhine-Westphalia)

Population: 16.7 million
Area: 34 071 km^2
Seat of Land government: Düsseldorf
Population: 570 000

Ministerium für Bauen und Wohnen, Postfach 10 11 03, D-40002
  Düsseldorf (*Tel:* +49-211 38 43-0; *fax:* +49-211 908 86 0)
VDZ Forschungsinstitut der Zementindustrie: Verein Deutscher
  Zementwerke eV (VDZ) (Research Institute of the Cement
  Industry), Postfach 30 10 63, Tannenstr. 2, Düsseldorf (*Tel:* +49-211
  457 8–1)
VDI-Gesellschaft Bautechnik (Building Technology Division of the
  Association of German Engineers), Graf-Recke-Strasse 84, Postfach
  1139, Düsseldorf 1 (*Tel:* +49-211 621 41)

Other large centres

Aachen Nordrhein-Westfalen
Population: 246 000
TU (Rheinisch-Westfälische Technische Hochschule), Templergraben
  55, Aachen
ibac Institut für Bauforschung (Institute for Building Research),
  Schinkelstr 3 (*Tel:* +49-241 80 51 00)
LBB Landesinstitut für Bauwesen und Bauschadensforschung
  (Institute for Building and Building Failures Research), Jakobstr 2
  (*Tel:* +49-241 422 22 7)
Lehrstuhl für Baustatik und Massivbau (Structural Engineering and
  Construction Department), Mies-van-der-Rohe-str.
Lehrstuhl für Hochbauentwurf und Industriebau (Building Design and
  Industrial Building Department); Lehrstuhl und Institut für
  Strassenwesen (Highway Engineering Department), Templergraben
  55 (*Tel:* +49-241 422 22 12)

Bergisch Gladbach Nordrhein-Westfalen [+49/2202]
Population: 102 100
Bundesanstalt für Strassenwesen (Federal Highway Institute), Postfach
  100150, Brüderstrasse 53, Bergisch Gladbach (*Tel:* +49-2202 4-43-0)

Bonn Nordrhein-Westfalen
Population 296 000
Rheinische Friedrich Wilhems Universität; Institut für Städtebau,
  Siedlungswesen und Kulturtechnik, Nussallee 1 (*Tel:* +49-228 73 26
  10)

Dortmund Nordrhein-Westfalen
Population: 570 000
MPA NRW Staatliches Materialprüfungsamt Nordrhein-Westfalen,
  Marsbruchstrasse 186, Dortmund 41-Aplerbeck (*Tel:* +49-231 4502-
  1)

Duisberg Nordrhein-Westfalen
Population: 540 000
FEhs Forschungsgemeinschaft Eisenbüttenschlacken, Bliersheimer
  Strasse 62, Duisberg 14 (*Tel:* +49-203 35-4 70 86)

Köln Nordrhein-Westfalen
Population: 107 000
Institut für Kalk- und Mörtel-Forschung eV. Annastrasse 67-71, Köln 51 (*Tel:* +49-221 370 400 44)

Munster Nordrhein-Westfalen
Population: 252 000
Westfälische Wilhems Universität, Munster
Institut für Siedlungs- und Wohnungwesen, Am Stadtgraben 9, Munster (*Tel:* +49-251 490 29 70)

Troisdorf Nordrhein-Westfalen
Population: 66 000
Institut für Baustoffprüfung und Fussbodenforschung (Institute for Testing of Building Materials and Flooring Research), Industriestrasse 19, Troisdorf (*Tel:* +49-2241 420 42)

Land Rheinland-Pfalz (Rhineland-Palatinate)

Population: 1.7 million
Area: 19 848 km²
Seat of **Land** *government:* Mainz
Population: 180 000
Innenministerium, Schillerplatz 3-5, D-55116
(*Tel:* +49-6131 16-0; *fax:* +49-6131 16-35 95)
Fachhochschule des Landes Rheinland-Pfalz, Holzstrasse 36
(*Tel:* +49-6131 3 27 25)

Other centres

Kaiserslautern
Population: 104 000
MPA Universität Kaiserslautern Materialprüfamt, Gottlieb-Daimler-Str. (*Tel:* +49-63 205 3003)

Siegen
Population: 119 000
Universität Gesamthochschule Siegen, Baustoffprüfstelle, Postfach 101240, Paul Boutz Strasse 9/11 (*Tel:* +49-271 74 01)

Saarland
Population: 1.0 million
Area: 2 576 km²
Seat of **Land** *government:* Saarbrücken
Population: 200,000
Innenministerium, Franz-Josef Röderstr. 21, Saarbrücken 1 D-55119
(*Tel:* +49-681 5 01-00; *fax:* +49-681 5 01-22 22)

Land Sachsen-Anhalt (Saxony-Anhalt)

Population: 2.9 million
Area: 20 455 km^2
Seat of **Land** *government:* Magdeburg
Population: 287 000
Ministerium für Raumordnung, Städtebau und Wohnungwesen, Herrenkrugstrasse 69, D-39114 Magdeburg (*Tel:* +49-391 5 67-75; *fax:* +49-391 5 67 75 10)

Other large centre

Halle (Saale) [+49/345 . . .]
Population: 230 000

Freistaat Sachsen (Saxony)

Population: 4.9 million
Area: 18 337 km^2
Seat of **Land** *government:* Dresden
Population: 500 000
Innenministerium, Archivstrasse 1, D-01097
(*Tel:* +49-351 5 64-0; *fax:* +49-351 59 17 32)
Bau-aussenstelle Dresden, Gerhart-Hauptmannstr. 1, Dresden
  (Federal building agency for reconstruction)

Other large centre

Leipzig [+49/341 . . .]
Population: 530 000

Land Schleswig-Holstein

Population: 2.8 million
Area: 15 729 km^2
Seat of **Land** *government:* Kiel
Population: 245 000
Innenministerium, Düsterbrookerweg 92, D-24105
(*Tel:* +49-431 5 96-1; *fax:* 49-431 5 99 3003)
Materialprüfanstalt Professor Matthaes, Hamburger Chaussée 94
(*Tel:* +49-431 68 68 99)

Other large centre

Lübeck [49/451 . . .]
Population: 216 000

Land Thüringen (Thuringia)

Population: 2.6 million
Area: 16 251 km²
Seat of **Land** *government:* Erfurt
Population: 220 000
Innenministerium, Schillerstr. 27, D-99096 Erfurt
(*Tel:* +49-361 3 98-0; *fax:* +49-361 3 98 21-28)

15.5 FINLAND

SF Finland: Suomen Tasavalta   **(EFTA Member)**
Population: 4 984 000 (15/km²); 60% urban
Land area: 338 145 km²
Capital city: Helsinki (490 034)
Standards body: SFS Suomen Standardisoimisliiro r.y.
Tel: +358 0 64 56 01; *fax:* +358 0 64 41 47
*UEAtc member:*VTT Valtion Teknillinen Tutkimuskesus (Technical Research Centre of Finland)
Tel: +358 45 61; *fax:* +358 45 67 017
CIB full member: VTT (as above)

Finnish Embassy: 38 Chesham Place, London SW1X 8HW, UK
Tel: 0171 838 6200

National language: Finish
Currency: Markka (Fmk)
Suomen Valhokalenteri – Finlands Statskalendar, published by Helsingfors Universitet, lists public bodies

Suomen Standardisoimisliitto r.y. (SFS)
PO Box 205, 00121 Helsinki 12
Tel: +358 0 64 56 01; *fax:* +358 0 64 31 47
National standards organization; member of ISO and CEN. Operates a SFS Mark licensing third-party certification scheme, quality system to SFS-ISO 9001-3 and acts as the Finnish agency of the Nordic Environmental Labelling Board. SFS publishes an annual *Standards Catalogue* in English and Finnish.

Finnish Ministry of the Environment
PO Box 306, SF-00531 Helsinki
Tel: +358 0 16 01
Responsible for RakMK Suomen rakennusmääräyskokoelma (National Building Code of Finland).

Valtion Teknillinen Tutkimuskeskus (VTT)
Vuorimiehentie 5, SF-02150 Espoo 15
Tel: +358 45 61; *fax:* 358 45 67 017
Division of building research, Technical Research centre of Finland, is the principal Finnish research and testing organization for construction, under-

taking tests and inspections for ministry of the environment directly or through industry laboratories under VTT control. Amendment to building law will give VTT the following roles: to be a notified testing body; to be designated as the Finnish body issuing European technical approvals when EFTA member states are accepted into EOTA; and, as member UEAtc, to continue to issue UEAtc *agrément*-type certificates.

International Department, Nordtest
Ostanvindsvagen 2, Postbox 111, SF-02101 Esbo

Rakennustietosäätiö (RTS)
PO Box 1004, SF-0010 Helsinki
Tel: +358 0 6944911; *fax:* +358 0 6931537
Finnish Building Information Institute; also operates an information and publishing service, and a bookshop.

Rakennusmestarien Kekuskuslitto by (RKL)
Ratiaka Marinpölti 3, SF 00240 Helsinki
Tel: +358 0 645601; *fax:* 33 0 656209
Finnish Association of Construction Engineers, founded 1905.

Rakennustaito
Rahakamarinportti 3A, SF-00240 Helsinki
Tel: +358 0 144 133; *fax:* +358 0 147 080
Magazine of Rakennusmestarien keskusliitto ry (Central Association of Construction Engineers).

In association with VTT, the Ministry of Environment operates a number of central approval schemes for construction products, modelled on those operated by Boverket in Sweden.

15.6   FRANCE

F France: La République Française   **(EC Member State)**
Population: 56 580 000 (103/km^2); 74% urban
Land area: 547 026 km^2
Capital city: Paris (2 078 900); Greater Paris (8 706 963)
Standards body: AFNOR Association française de normalisation
Tel: +33 1 42 91 55 55; *fax:* +33 1 42 91 56 56
ETA Bodies: CSTB Centre Scientifique et Technique du Bâtiment (spokesbody on EOTA)
Tel: +33 1 40 50 28 28; *fax:* +33 1 45 25 61 51
SETRA Service d'Études Techniques des Routes et Autoroutes
Tel: +33 1 46 34 49 27; *fax:* +33 1 46 31 33 69
UEAtc member: CSTB (as above)
CIB full member: CSTB (as above), UTI

French Embassy: 58 Knightsbridge, London SW1X 7JT, UK
Tel: 0171 201 1000

National language: French
Currency: French franc (Ffr)

Bottin Administratif, published by Bottin, 31 Cours de Juilliottes, F-94706 Maison-Alfort Cedex (*Tel:* +33 1 49 81 56 56) gives information on government departments and agencies.

Association française de normalisation (AFNOR)
Tour Europe, Cedex 07, F-92080 Paris La Defense
Tel: +33 1 42 91 55 55; *fax:* +33 1 42 91 56 56
National standards association, representing France in ISO and other international standards organisations. AFNOR operates some 70 certification schemes for building products as an *organisme certificateur agréé* certifying conformance to French normes for NF-Mark; also operates a *NF-Réaction au feu* reaction-to-fire classification scheme. Mainly for heating and domestic hot-water installations and air-conditioning, products assessed against standards not officially recognized (*normes non-homologuées*) carry NF-Marque Certinor. Certificate 3AQ indicates that procedures for quality management assessment and certification based on ISO 9000 have been followed. Full list is on Promolog database; annual *AFNOR Catalogue* list over 4500 official English translations, mostly prepared with BSI THE.

Ministère Équipement, Transport, Logement (MELT)
Pilier Nord, Arche Défense, F-92055 Paris-La-Défense Cedex 04
Tel: +33 1 40 81 21 22; *fax:* +33 1 40 81 27 81
French construction, housing and transport ministry.

Ministère de l'Industrie
99-101 rue de Grenelle, F-75007 Paris Cedex 07
Tel: +33 1 45 56 36 36; *fax:* +33 1 45 56 34 94
French ministry of trade and industry, responsible for AFNOR.

Centre Scientifique et Technique du Bâtiment (CSTB)
4 avenue du Recteur Poincaré. F-75782 Paris Cedex 16
Tel: +33 1 40 50 28 28; *fax:* +33 1 45 25 61 51

Laboratories:
84 avenue Jean Jaurès, Boite Postale 2, F-77421 Marne-la-Vallée Cedex 2
Tel: +33 1-64 68 82 82; *fax:* +33 1-64 63 84 49
Grenoble, 24 rue Joseph-Fourier, F-38400 St-Martin-d'Hères
Tel: +33 16 76 54 11 63
Nantes, 11 rue Henri-Picherit, F-44300 Nantes
Tel: +33 16 40 59 42 55
Sophia-Antipolis, BP 21, F-06562 Valbonne Cedex
Tel: +33 16 93 74 63 63
Government-supported centre for construction research – analogous to UK's BRE; also responsible for administration and technical support for *avis technique*; DTU, the nearest equivalent to UK codes of practice; and Section of the NOREX export promotion system, devoted to assisting construction.

Service d'Études Techniques des Routes et Autoroutes (SETRA)
BP 120, 46 avenue Aristide Briand, F-92223 Bagneux
Tel: +33 1 42 31 33 70; *fax:* +33 1 42 31 33 69
Technical research centre under MELT for highway design and construction.

Laboratoire Central des Ponts et Chaussées (LCPC)
58 bde. Lefebfre, Cedex 15, F-75732 Paris
Tel: +33 1-40 43 50 10; *fax:* +33 1-40 19 49 11
Central civil engineering laboratory of MELT (Ministry of construction and transport); there is a second establishment at Bouguenais, Nantes.

Centre Technique du Bois et de l'Ameublement (CTBA)
10 avenue de Saint Mandé, F-75012 Paris
Tel: +33 1-40 19 49 10; *fax:* +33 1-40 10 49 11
French technical and research centre for timber, timber products and furniture: *organisme certificateur agrée* for timber and timber products, including timber-frame housing components: responsible for NF Reaction to Fire schemes for panel and sheet materials: with wood preservation association (AFBS) manages certification scheme for timber treatment. Through EURODOC, provides extensive information and publication service, including *La Normalisation et l'Europe* covering normes, product certification; certification of firms – in timber and timber products.

Laboratoire Central des Industries Electriques (LCIE)
33 avenue du Général Leclerc, F-92260 Fontenay aux Roses.
Tel: +33 1-46 45 21 84
Central laboratory of French electricity industry.

Laboratoire National d'Essais (LNE)
1 rue Gaston Boissier, F-75015 Paris
Tel: +33 1-45 32 29 89
French national testing laboratory.

Institut des Sciences et des Techniques de l'Équipement et l'Environment pour le Dévéloppement (ISTED)
38 rue Liancourt, F-75014 Paris
Tel: +33 1 433 556 67
Members include CEBTP, CSTB, ENPC, LCPC and SETRA.

Association Française de Recherches et d'Essais sur les Matériaux et les Construction (AFREM)
Domaine de Saint-Paul, BP 37, F-78470 Saint-Rémy-les-Chevreuse
Tel: +33 1-30 85 20 00; *fax:* +33 1-30 85 20 47
Association of French construction materials laboratories.

Agence pour la Prévention des Désordres et l'Amélioration de la Qualité de la Construction (Qualité Construction)
30 Place de la Madeleine, F-75008 Paris
Tel: +33 1-42 65 44 32

Association pour l'Étude de la Pathologie et de l'Entretien du Bâtiment (EPEBAT-CSTB)
84 avenue Jean-Jaurès, Champs-sur-Marne, BP 02, F-77421 Champs-sur-Marne
Tel: +33 1-60 05 90 58
Association set up jointly by CSTB and insurers to study and prevent building defects.

AFCIQ Association Française pour la Qualité
Tour Europe, Cedex 7, F-92080 Paris La Defense
Tel: +33 1-42 91 59 53; *fax:* +33 1-42 91 56 56

Assemblé Plénière des Sociéties d'Assurances contre l'Incendie et les Risques Divers (APSAIRD)
11 rue Pillet-Will, F-75009 Paris
Tel: +33 1-47 46 82 49
Certification body for security and fire detection and alarm systems.

Association technique de l'industrie du gaz en France (ATG)
62 rue de Courcelles, F-75008 Paris
Tel: +33 1-47 54 20 20; *fax:* 1 42 27 49 43
Technical association of French gas industry.

Confédération de l'Artisanat et des Petites Entreprises du Bâtiment (CAPEB)
2 bis rue Michelet, F-92133 Issy-les-Moulineaux Cedex
Tel: +33 1-45 54 95 60
Representative organization of French small artisan builders.

Centre de Documentation Scientifique et Technique de CNRS (CD CNRS)
26 rue Boyer, F-75020 Paris
Tel: +33 1 43 58 35 59
Documentation centre of French national scientific research council.

Centre Expérimental de Recherches et d'Etudes du Bâtiment et des Travaux Publics (CEBTP)
Domaine de Saint-Paul, BP 1, F-78470 Saint-Rémy-les-Chevreuse
Tel: +33 3-05 29 200
The building and civil engineering laboratory component of UTI.

Comité Scientifique et Technique de l'Industrie, de Chauffage, de la Ventilation et du Conditionnement d'Air (COSTIC)
Domaine de Saint-Paul, BP 37, F-78470 Saint-Rémy-les-Chevreuse
Tel: +33 1-30 85 20 10; *fax:* +33 1-30 85 20 38
French heating, air conditioning and ventilation industry's research association.

Fédération Nationale du Bâtiment (FNB)
33 Av Kleber, F-75784 Paris Cedex 16
Tel: +33 40 69 51 00
National federation of major building contractors.

FNTCP Fédération Nationale des Travaux Publics (FNTP)
3 rue de Berri, F-75008 Paris
Tel: +33 1-45 63 11 44
National federation of civil engineering and public works contractors.

Le Moniteur des Travaux Publics et du Bâtiment
17 rue d'Uzes, F-75002 Paris Cedex 02
Tel: +33 1-42 96 15 50

Principal weekly building and civil engineering magazine; reports texts of all French government legal instruments, standards, *avis technique* etc. relevant to construction in weekly supplement.

Organisme professionnel de qualification et classification du bâtiment et des activitiés (OPQCB)
7 rue La Pérouse, F-5784 Paris Cedex 26
Tel: +33 1-45 53 55 30
Maintains classified register of building firms and associated specialists.

Organisme professionnel de qualification des ingénieurs–conseils et bureau d'études techniques du bâtiment et des infrastructures (OPQIBI)
Maison de l'Ingénierie, 3 rue Léon-Bonnat, F-75016 Paris
Tel: +33 1-45 67 35 34
Maintains register of consulting engineers and technical design offices.

Union Nationale des Syndicats Français des Architectes (UNSFA)
26 bde Raspail, F-75007 Paris
Tel: +33 1 45 44 58 45
Federation of associations of French architects.

Union Technique Interprofessionelle des Fédérations Nationales du Bâtiment et des Travaux Publics (UTI FNB)
9 rue La Pérouse, F-75784 Paris Cedex 16
Tel: +33 1-47 20 10 20
French contractors federations' technical service. Two of its principal components are based at the Saint-Rémy-les-Chevreuse site: CACT (Centre d'Assistance au Calcu Technique), directed at use of computers in construction; and CATED (Centre d'Assistance Technique et Documentation), advisory and information service.

Licensed technical control offices:

Contrôle Construction (AINF)
9–11 rue Georges Enesco, F-94008 Creteil
APAVE, 102 rue des Poissoniers, F-75018 Paris
Tel: +33 1-42 57 11 05

Bureau VERITAS
17 bis place Reflets, La Défense, F-92400 Courbevoie
Tel: +33 1-42 91 52 91

Contrôle et Prévention (CEP)
34 rue Rennequin, F-75017 Paris
Tel: +33 1-47 66 52 72

SOCOTEC SA Headquarters
Tour Maine-Montparnasse, 33 avenue du Maine, F-75755 Paris Cedex 15
Tel: +33 1-45 38 52 73

Direction des Services Techniques
3 avenue du Centre, F-78182 Saint-Quentin-en-Yvelines Cedex

Association SOCOTEC Qualité
3 avenue du Centre, F-78182 Saint-Quentin-en-Yvelines
Tel: +33 30 43 99 13

Comité des organismes de prévention et de contrôle technique (COPREC)
33 avenue du Maine, 75755 Paris Cedex 15
Professional association of licensed technical control offices.

For an account of construction product certification arrangements, see *Building Technical File 23* (October 1988), and for the French technical control system see *Building Technical File 27* (October 1989).
  CIRIA Special Publication 66 (1989), *The French construction industry: A guide for UK professionals*, describes the background, market, legal and institutional framework, and the French construction industry.

15.7 GREECE

GR Greece: Elliniki Dimokratia (EC Member State)
Population: 10 075 000 (76/km^2)
Land area: 131 990 km^2
Capital city: Athens (855 337)
Standards body: ELOT Ellinikos Organismos Typopoiisis
Tel: +30 1 201 50 25; *fax:* +30 1 202 38 25
ETA body: ELOT (as above)
UEAtc member: none
CIB full member: none

Embassy of Greece; 1A Holland Park, London W11 3TP, UK
Tel: 017 1 229 3850

National language: Greek
Currency: drachma

Ellinikos Organismos Typopoiisis (ELOT) (Hellenic Organization for Standardization)
313 Acharon Street, GR-11145 Athens
Tel: +30 1 201 50 25; *fax:* +30 1 202 38 25
National standards organization; only body in Greece legally authorized to undertake certification procedures and grant conformity marks; designate Greek technical approval body under 89/106/EEC; undertakes testing in its own laboratory – there are also testing facilities at the National Technical University of Athens.

Ministry of the Environment, Public Works and Urban Planning
Panormou 2, EL - 11523 Athens
Tel: +30 1 644 26 82; *fax:* +30 1 643 44 70

Ministry of Industry, Standardization Section
Messogion 14-18, EL-1150 Athens
Tel: +30 1 778 17 31; *fax:* +30 1 779 88 90
Central Public Works Laboratory, 166 Piraeus Street, Athens
Tel: +30 1 345 59 21

Laboratory of Physical Chemistry and Applied Electochemistry
National Technical University of Athens, Departments of Theoretical
and Applied Mechanics, 5 Heroes of Polytechnion Avenue, GR-Athens
624
Tel: +30 1 779 55 75

Technical Chamber of Greece (TEE)
4 Karageorgi Servias Str, PO Box 673, GR-Athens 125
Tel: +30 1-325 45 91-9
The principal professional engineering organization, with responsibilities for
granting licenes to practice for architects and engineers.

15.8 IRELAND

IR Republic of Ireland:
Poblacht Na L'Eireann   **(EC Member State)**
Population: 3 471 000 (49/km^2)
Land area: 70 285 km^2
Capital city: Dublin (502 749)
Standards body: NSAI/EOLAS National Standards Authority of
Ireland
Tel: +353 1 37 01 01; *fax:* +353 1 36 98 21
ETA body: IAB Irish Agrément Board, ELOAS-NASI
Tel: +353 1 37 01 01; *fax:* +353 1 36 98 21
UEAtc member: IAB (as above)
CIB full member: none

Irish Embassy: 17 Grosvenor Place, London SW1 7HR, UK
Tel: 0171 235 2171

National languages: Erse; English
Currency: Punt (IR£)

National Standards Authority of Ireland (NSAI/EOLAS)
Ballymun Road, Glasnevin, Dublin 9
Tel: +353 1 37 01 01; *fax:* +353 1 36 98 21
As well as being the Irish national standards body and responsible for IAB
and ILAB (Irish Laboratory Accreditation Board), NSAI/EOLAS replaced
IIRS (Institute for Industrial Research and Standards) as the Irish Science
and Technology Agency.

Irish Agrément Board (IAB)
NSAI/EOLAS National Standards Authority of Ireland, Ballymun Road, Glasnevin, Dublin 9
Tel: +353 1 37 01 01; *fax:* +353 1 36 98 21

Department of the Environment
Custom House, IRL-Dublin 1
Tel: +353 1 679 33 77; *fax:* +353 1 77 92 78
Central government department with responsibilities for physical and emergency planning, building control and housing; shares with NSAI representative of Irish Republic in implementation of 89/106/EEC.

Office of Public Works
Baile Atha Cliath 2, Dublin 2

Institution of Engineers of Ireland
22 Clyde Road, Balisbridge, Dublin 4
Tel: +353 1 684 341
There is an account of construction product certification arrangements in *Building Technical File 27* (October 1989).

15.9   ITALY

I Italy: Repubblica Italiana   (**EC Member State**)
Population: 57 630 000 (191 km^2); 68% urban
Land area: 301 277 km^2
Capital city: Rome (2 815 457)
Other large centres: Milan (1.5 million); Naples (1.2 million); Turin (1 million); Genoa and Palermo (over 0.7 million); Bologna and Florence (each over 0.4 million).
Standards body: UNI Ente Nazionale Italiano di Unificazione
Tel: +39 2 70 02 41; *fax:* +39 2 70 10 61 06
ETA body: STC Servizio Tecnico Centrale dei Consiglio Superiore dei Lavori Pubblici
Tel: +39 6 84 82 41 01; *fax:* +39 6 86 96 48
Centro Studi ed Esperienze Antincendio de Ministero dell'Interno
Tel: +39.6 718 77 19
ICITE Istituto Centrale per l'Industrializzione e la Tecnologia Edilizia
Tel: +39 2 98 061; *fax:* +39 2 98 28 00 88
UEAtc member: ICITE (as above)
CIB full member: ICITE (as above), IRIS

Italian Embassy: 14 Three Kings Yard, Davies Street, London W1Y 2 EH, UK
Tel: 0171 629 8200

Republic with some construction responsibilities, including planning and building regulation with regional and provincial governments.

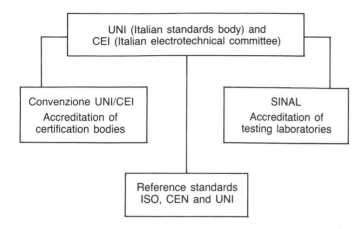

Figure 15.1 Italian arrangements for certification

National language: Italian; German is also spoken in Alo Adige
Currency: Lira (L)

Ente Nazionale Italiano di Unificazione (UNI)
Piazza A. Diaz 2, I-20123 Milano MI
Tel: +39 2 72 147; *fax:* +39 2 86 92 120
National standards body; member of ISO and CEN; with CEI Italian
Electrotechnical Committee under tutelage of Industry Ministry; participates
in accreditation of testing laboratories and certification bodies through SINAL;
with CEI manages SINCERT for the accreditation of certification bodies
against ISO 2, IEC 28, EN 45011 12, and with the ISO 9000 series. With CEI
issues monthly newsletter in Italian, with English and French summaries.

Sistema Nazionale per l'Accreditamento di Laborati (SINAL)
via Campania N.31., I-00187 Roma
Tel: +39 6 46 27 31
Founded in 1988 on the initiative of UNI and CEI, under the auspices of the
Ministry of Industry, Commerce and Artisanate, CNR Consiglio Nazionale
delle Ricerche (National Research Council) and other national organizations,
as a non-profit association for accreditation of Italian and foreign testing laboratories against the EN 45001 series of standards.

Istituto Centrale per l'Industrializzione e la Tecnologia Edilizia (ICITE)
Via Lombardia no 49 - Frazione Sesto Ulteriano, I-2009998 San
Guiliano Milanese (MI) Milano
Tel: +39 2 98 06 450; *fax:* +39 2 98 81 763
Set up in 1960 by CNR as independent construction research institute with
responsibilities for *agrément* work; full member of UEAtc and CIB member;
one of three designated Italian technical approval bodies under 89/106/EEC;
manages Italian quality assurance scheme for cements. Issues a quarterly
bulletin in Italian.

Centro Studi ed Esperienze del Corpo Nazionele dei Vigli del Fuoco
Piazza Seilla 2 - Caponelle, I-00178 Roma
Tel: +39.6 718 89 96; *fax:* +39.6 718 77 19
Fire testing centre of the nation's fire service; one of three designated technical approval bodies under the EU CPD; centre's certificates are recognized by the Interior Ministry under fire safety regulations.

Servizio Tecnico Centrale della Presidenza dei Consiglio Superiore dei Lavori Pubblici (STC)
via Nometana 2, I-00161 Roma
Tel: +39 6 844 18 79; *fax:* +39 6 855 96 48
Ministry of Public Works technical centre; one of three designated technical approval bodies under the EU CPD; under the 1975 law, approves a number of materials testing laboratories – private, 'official' and university – to issue certificates for cement, aggregates, concrete, reinforcement and structural steel, the Ministry's provincial directorates being responsible for laboratory surveillance.

Ministero dell Industria
via Molise 19, I-00187 Roma
Tel: +39 6 48 27 821; *fax:* +39 6 47 44 430

Ministero dell Interno
Piazza Scillia 3, I-00100 Roma
Tel: +39 6 71 80 55; *fax:* +39 6 71 87 719

Ministerio dei Lavori Pubblici
via Nomentana 2, I-00100 Roma
Tel: +39 6 84 82 31 14; *fax:* +39 6 86 71 87

Consiglio Nazionale delle Ricerche (CNR)
Ufficio Relazioni Internaxionali, Piazzale delle Science 7, I-00100 Roma
Tel: +39 6-09 59 345
International relations office of Italian national research council.

Istituto di Ricerche sul Legno
Piazza T A Edison 11, I-50133 Firenze
Tel: +39 55-570210

Istituto pe la Tecnologia del Legno
via Biasi 75, I-38010 S Michele a A Trento
Tel: +39 461 65 01 68
CNR timber and timber products research institute.

Isituto Italiano di Garanzia della Qualitåa per i Prodotti Siderurgici
piazza Velasca 8, I-20122 Milano
Tel: +39 2 870 651; *fax:* +39 2 805 7815
Certification body for steel products, including reinforcing steels.

Centro di Richercha e Sperimentazione per l'Industria Ceramica
via Martelli 26, I-40138 Bologna
Tel: +39 51-53 40 15
Principal ceramic products research institute.

Costruire
Corso Monforte 15, I-20122 Milano
Tel: +39 2 760 042 51; *fax:* +39 2 791 904
Leading monthly construction magazine.

For an account of construction product certification arrangements, see
Building Technical File 29 (April 1990)

15.10   LUXEMBOURG

L Luxembourg:
Grand-Duché de Luxembourg   **(EC Member State)**
Population: 379 000 (147/km²)
Land area: 2586 km²
Capital city: Luxembourg (76 10)
Standards body: ITM Inspection du Travail et Mines (Luxembourg)
Tel: +352 49 921 21 06; *fax:* +352 49 14 47
ETA body: Laboratoire des Ponts et Chaussées
Tel: +352 45 20 75; *fax:* +352 45 57 37
UEAtc member: none, see Belgium
CIB full member: none

Embassy: 27 Wilton Crescent, London SW1X 8SD, UK
Tel: 0171 235 6961

National language: French
Currency: Luxembourg franc (LFr)

Administration et Laboratoire des Ponts et Chaussées
7–15 rue Albert 1er, L-1117 Luxembourg
Tel: +352 45 20 75; *fax:* +352 45 49 61 03
Public works laboratory; designated Luxembourg technical approval body
under the EU Construction Products Directive and member of EOTA; tests
on and issues certificates for a limited list of materials based on technical
clauses in Luxembourg official specifications, or recognized standards from
neighbouring countries.

Chambre des Métiers
41 av. Glesner, L-1631 Luxembourg
Tel: +352 40 00 221; *fax:* +352 49 23 80
Smallest of the European Union Member States, Luxembourg does not have
its own *agrément* body or national standards organization; construction standards and certification practices usually follow those of Belgium.

15.11 NETHERLANDS

NL Netherlands:
Koninkrijk der Kederlanden **(EC Member State)**
Population: 14 980 000 (359/km²); 88% urban
Land area: 41 785 km²
Capital city: The Hague (443 900)
Standards body: NNI Nederlands Normalisatie-instituut
Tel: +31 15 69 03 90; *fax:* +31 15 69 01 90
ETA bodies: SBK (see below), BDA Keurings- en Certificerings-instituut BV
Tel: +31 1830 69 63 6; *fax:* +31 1830 60 86 6
BKB BV Kwaliteitsverklaringen Bouw
Tel: +31 10 430 92 73; *fax:* +31 10 413 27 69
Stitchting BMC
Tel: +31 1820 39 23 3; *fax:* +31 1820 39 72 6
IKOB
Stichting Instituut voor Keuring en Onderzoek van Bouwmaterialen
Tel: +31 3420 13 64 3; *fax:* +31 3420 93 13 6
INTRON
Instituut voor materiaal- en milieu-onderzoek BV
Tel: +31 3403 795 80; *fax:* +31 3403 796 80
KIWA
Keuringsinstituut voor Waterleidingartikelen NV
Tel: +31 70 95 35 35; *fax:* +31 70 95 34 20
SKG
Stichting Kwaliteitscentrum Gevelelementen
Tel: +31 79 51 18 03; *fax:* +31 79 53 13 65
SKH
Stichting Keuringsbureau Hout
Tel: +31 2152 68 73 7; *fax:* +31 2152 68 38 1
UEAtc member: SBK Stichting Bouwkwaliteit
Tel: +31 70 399 84 67; *fax:* +31 70 390 29 47
CIB full members: BC Stichting Bouwcentrum
Tel: +31 10 30 99 11
SBR Stichting Bouwresearch
Tel: +31 10 12 35 28
TNO-Bouw
Tel: +31 15 84 22 68; *fax:* +31 15 84 39 90
VROM
Tel: +31 79 27 91 11

Royal Netherlands Embassy: 38 Hyde Park Gate, London SW7 5DP, UK
Tel: 0171 584 5040

National language: Dutch
Currency: Florin (N Fl)

Netherlands Normalisatie-instituut (NNI)
Kalfjeslaan 2, PO Box 5059, NL-2600 GB Delft
Tel: +31 15 69 03 90; *fax:* +31 15 69 01 90
The Netherlands standards organization – not a certification body.

Stichting voor Onderzoek, Voorschriften en Kwaliteitseisen op het Gebied van Beton (CUR-VB)
PO Box 61, Bredewater 26, 2700 AB Zoetermeer
Tel: +31 79 219 31 3
Netherlands Committee for Research, Codes and Specifications for Concrete; formed from 1979 amalgamation of CUR and CVB (Committee on Codes for Concrete). CUR-VB Codes published by NNI as NENs (Dutch Standards); or if not definitive as CUR-VB Recommendations.

Ministerie van Volkshuisvesting, Riumtelijke Ordening en Milieubeheer (VROM)
PO Box 20951, Van Alkemadelaan 85, NL-2500 EZ Den Haag
Tel: +31 70 335 35 35
Directie Coordinatie Bouwbeleid, PO Box 3001, Boerhaavelaan 5, NL-2700 KA Zoetermeer
Tel: +31 79 27 91 11
Directorate of building policy, with responsibility for building legislation. VROM is the Dutch Ministry of housing, planning and construction; its head-quarters are in the Hague, and most of its technical directorates in the new town of Zoetermeer. With the trade ministry, VROM is responsible for implementing CPD through directorates-general for housing and research and quality assurance.

Ministertie Economische Zaken (EZ)
PO Box 20101, NL-2500 EC Den Haag
Ministry for Economic Affairs, shares with VROM responsibilities for implementing 89/106/EEC.

TNO-Bouw
PO Box 49, NL-2600 AA Delft (Lange Kieweg 5, Rijswijk)
Tel: +31 15 84 20 00; *fax:* +31 15 84 39 90
TNO Institute for Building and Construction Research; technical centre for fire prevention; preparing performance-based building regulations for VROM.

Raad voor Certificatie (RvC)
Stationsweg 11, NL-3972 KA Driebergen
Tel: +31 3438 12604
Dutch council for certification, responsible for the accreditation of certification bodies, including STERLAB (Nederlandse Stichting voor Erkenning van Laboratoria), Certification body for testing laboratories; has cooperative arrangements with UK NACCB.

Stichting Bouwkwaliteir (SBK)
Cobbenhage, Treubstraat 1, NL-2288 EG Rijswijk ZH
Tel: +31 70 399 84 67, fax +31 70 390 29 47
Foundation for Building Quality; Dutch EOTA spokesbody, although not an ETA body; has close links with RvC, advising the council on certification

matters relating to construction; issues quarterly in Dutch, *Bouwkwaliteit-Nieuws*.

Vereniging van Certificatie-Instellingen in de Bouw in Nederland (VECIBIN)

PO Box 225, NL-2280 AE Rijswijk, Sir Winston Churchill-Laan 273, NL-2288 EA Rijswijk

Tel: +31 70 395 36 05; *fax:* +31 70 395 34 20

Association of certification bodies for building materials and components in the Netherlands; an active member of the joint RvC/SBK harmonization committee for the building section; with SBK publishes an annual guide to certificated construction products, *Kwaliteitsverklaringengids*.

Keurings- en Certificeringsinstituut BV (BDA)

Avelingen West 35 PO Box 739, NL-4200 AS Gorinchem

Tel: +31 1830 6 96 27; *fax:* +31 1830 6 08 66

Dutch ETA body; member of VECIBIN. BDA's work is in the field of installation of roof-covering systems and materials, board and slab materials, polyurethane foam, polyisocyanate foam, expanded polystyrene foam, mineral wool, expanded perlite and roofing membranes.

BV Kwaliteitsverklaringen Bouw (BKB)

Weena 760, PO Box 1836, NL-3000 BV Rotterdam

Tel: +31 10 430 92 73; *fax:* +31 10 413 27 69

Dutch ETA body; member of VECIBIN; jointly responsible with TNo for a number of certification schemes including cavity fill, building systems.

Stichting BMC

Bleulandweg 1b, PO Box 150, Nl-2800 AD Gouda

Tel: +31 1820 3 23 00; *fax:* +31 1820 7 02 16

Dutch ETA body; member of VECIBIN. Set up in 1962 by the Dutch Concrete Association, BMC operates certification schemes for concrete products.

Stichting Instituut voor Keuring en Onderzoek van Bouwmaterialen (IKOB)

Ambachtsweg 10, PO Box 205, NL-3770 AE Barneveld

Tel: +31 3420 13 64 3; *fax:* +31 3420 93 13 6

Dutch ETA body; member of VECIBIN. Set up in 1983, IKOB is a non-profit-making foundation serving the brick and aerated concrete products industries.

Instituut voor materiaal- en milieu-onderzoek BV, afdeling INTRON-certificatie (INTRON)

Onderdoor 19, PO Box 226, NL-3990 GA Houten

Tel: +31 3403 795 80; *fax:* +31 3403 796 80

Dutch ETA body; member of VECIBIN. Set up in 1971, INTRON is responsible for certification of concrete repair firms.

KIWA NV Certificatie en Keuringen

Sir Winston Churchill-laan 273, Postbus 70, NL-2280 AE Rijswijk

Tel: +31 70 395 35 35; *fax:* +31 70 395 34 20

Dutch ETA body; member of VECIBIN, KIWA, the testing and inspection organization of the Dutch water authorities, is the principal certification body

serving the construction industry. It issues KIWA-*attests* of an *agrément* type
to specifications indicating specific uses as well as conformity to standards
certificates.

Stichting Kwaliteitscentrum Gevelelementen (SKG)
Veldzigt 30B, Postbus 212, NL-3454 ZL De Meen
Tel: +31 3406 2 16 33; *fax:* +31 3406 2 16 33 77
Dutch ETA body; member of VECIBIN. Set up in 1977, SKG is a non-
profitmaking foundation working on windows and curtain walling.

Stichting Keuringsbureau Hout (SKH)
Huizermaatweg 29, PO Box 50, NL-1270 AB Huizen
Tel: +31 2152 68 73 7; *fax:* +31 2152 68 31 1
Dutch ETA body; member of VECIBIN. Set up in 1962, SKH is a non-profit-
making foundation accredited in the field of wood and wood products.

VEG Gasinstituut
PO Box 137, Wilhemsdorf 50, NL-7300 AC Apeldoorn
Member of VECIBIN; product and quality system certification for gas
appliances.

TNO–Organisatie voor Toegepast Natuurwetenschappelijk Onderzoek
Central Organisatie TNO, Juliana van Stolberglaan 148, PO Box 297,
NL-2501 BD 's-Gravenhage
Tel: +31 15 13 82 22

Netherlands Organization for Applied Scientific Research
TNO Research Institute for Environmental Hygiene
PO Box 214, NL-2600 AE Delft
Tel: +31 15 56 93 30

Nederlandse Stichting voor de Erkenning van Laboratorie (STERLAB)
Postbus 38, NL-2600 Delft
Accreditation of testing laboratories.

Pyttersen's *Nederlandse Almanak*, published by Bohn Stafleu Van
Loghem, PO Box 246, NL-3990 GA Houten, lists government and
other public bodies. For construction product certification arrangements see *Building Technical Files 26* (July 1989) and *31* (October
1990). Al Alderson, *The Netherlands: Product approval, quality control
and building control procedures*, BRE Report (1974).

15.12   NORWAY

N Norway: Kongeriket Norge   **(EFTA Member)**
Population: 4 721 000 (11/km^2); 74% urban
Land area: 386 975 km^2
Capital city: Oslo (452 415)

Standards body: NSF Norges Standardiseringsforbund
Tel: +47 2 46 60 94; ***fax:*** +47 2 46 44 77
UEAtc member: NBI Norges byggforskningsinstitutt
(Norwegian Building Research Institute)
Tel: +47 2 96 56 00; ***fax:*** +47 2 69 94 38
CIB full member: NBI (as above)

Royal Norwegian Embassy: 28 Belgrave Square, London SW1X
8QD, UK
Tel: 0171 235 7151

National language: Norwegian;
Currency: Norwegian krone (Nkr)

Norges Standardiseringsforbund (NSF)
PO Box 7026, 8-16 Homansbyen, N-0306 Oslo
Tel: +47 2 46 60 94; ***fax:*** +47 2 46 44 77
Norwegian standards organization; member of ISO and CEN.

Norges Byggstandardiseringrad
PO Box 129, Lobenhavngata 10, N-0566 Oslo
Tel: +47 2 35 50 20; ***fax:*** +47 2 35 50 76
Norwegian council for building standardization.

Norges Byggforskningsinstitutt (NBI)
Forskningsveien 3b, PO Box 123, Blindern, N-0314 Oslo 3
Tel: +47 2 96 55 00; ***fax:*** +47 2 69 94 38
Norwegian Building Research Institute; member of CIB and UEAtc.

Norsk Byggtjeneste A/s (NBT)
PO Box 1575,. Haakon VII's Gate 5VI, N-0118 Oslo 1
Tel: +47 2 83 3690; ***fax:*** +47 2 83 4233
Norwegian Building Centre providing publishing and information services, and
operates a bookshop. *Norsk Byggkatalog* (Norwegian Building Catalogue).

Norges Betongindustriforbund (NBIF)
PO Box 70, Kjelsåveien 150, Kjelsås, N-Oslo 4
Tel: +47 2 15 84 50
Norwegian Precast Concrete Foundation.

Norges Geotekniske Instituut (NGI)
Sognsveien 72, N-Oslo 8
Tel: +47 2 23 0388
Norwegian Geotechnical Institute.

Norsk Treteknisk Instituut (NTI)
Forskningsveien 3B, PO Box 322, Blindern, N-0314 Oslo 3
Tel: +47 2 46 98 80; ***fax:*** +47 2 9 94 38
Norwegian Institute of Wood Technology.

Arkitektur og byggteknik (SINTEF)
Alfred Getz veien 3, n-7034 Trondheim-Nth
Tel: +47 7 59 26 20

Division of Architecture and Building Technology, Trondheim University of Technology.

Norges Branntekniske Laboratorium (SINTEF)
N-7034 Trondheim-Nth
Tel: +47 7 50 41 90
Norwegian Fire Research Laboratory.

Akustisk Laboratorium/ELAB (SINTEF)
N-7034 Trondheim Nth
Tel: +47 7 59 26 36
Acoustic Laboratory at the Norwegian Institute of Technology.

Statens teknologiske instituut, Bygg- og anleggsavdelingen (STI)
Akersvn 24C, N-8116 Oslo 1
Tel: +47 2 20 45 50
Building and Civil Engineering Department, National Institute of Technology.

Statens Byggingstekniske Etat
PO Box 8185, N-0034 Oslo 1
Tel: +47 2 20 80 15
National Office of Building Technology and Administration, with responsibilities for building regulations, including fire protection and safety.

Byggeindustriens Landsforening
Oisbiks 112, Blindern, N-0314 Oslo 3
Tel: +47 2 95 55 00; *fax:* +47 2 95 55 23

Institute of Private Law
University of Oslo, Karl Johansgt 47, N-0162 Oslo
Tel: +47 2 42 90 10; *fax:* +47 2 42 69 57

Byggeindustrien
Holtegaten 26, N-0355 Oslo 3
Tel: +47 2 46 18 54; *fax:* +47 2 59 58 56
Norway's principal construction magazine.

Norwegian Building Regulations 1987
NS 3430: 1992 General conditions for contracts concerning construction and building
O. Sjolt, *Quality Assurance in Norway*, CIB Publication 109, 1989

15.13 PORTUGAL

P Portugal: Republica Portuguesa   **(EC Member State)**
Population: 10 560 000 (115/km²), 32% urban
Land area: 91 985 km²
Capital city: Lisbon (807 167)
Standards body: IPQ Instituto Portgués da Qualidade
Tel: +35 1 52 39 78; *fax:* +35 1 53 00 33
ETA body: LNEC Laboratório Nacional de Engenharia Civil

Tel: +351 1 84 82 131; *fax:* +351 1 89 76 60
UEAtc member: LNEC (as above)
CIB full member: LNEC (as above)

Portuguese Embassy: 11 Belgrave Square, London SW1X 8PP, UK
Tel: 0171 235 5331

National language: Portuguese
Currency: Escudo (Esc)

Instituto Portgués da Qualidade (IPQ)
Rua José Estévao 83A, P-1199 Lisboa CODEX
Tel +351 1 52 39 78; *fax:* +351 1 53 00 33
Since 1983, national standards organization; member of ISO and CEN; responsible for accreditation of certification and inspection bodies and testing laboratories.

Ministerio das Obras Publicas Transportes e Communicaçoes (MOPTC)
Praça do Comércio, P-1100 Lisboa
Tel: +351 1 87 95 41
Ministry of Public Works and Transport, whose overall responsibilities include technical building regulations, housing, telecommunications and urban transport – in the case of building standards and regulations through a Commission for the Revision and Establishment of Technical Regulations, and for roads through a national agency, JAE (Junta Autonoma das Stradas).

Laboratório Nacional de Engenharia Civil (LNEC)
Avenida do Brazil 101, P-1799 Lisboa CODEX
Tel: +351 1 84 82 131; *fax:* +351 1 89 76 60
National Civil Engineering Laboratory of Portugal, under MOPTC (Ministério das Obras Publicas, Transportes e Communicaçoes): designated Portuguese technical approval body under the EU CPD, and member of EOTA; founder member of UEAtc. Under technical building regulations, non-traditional products must have an LNEC agrément.

Instituto dos Produtos Florestais
Rua Filipe Falque 10-1, Apartado 1134, P-1003 Lisboa CODEX
Tel: +351 1 55 91 17
Timber products Institute under Ministry of Commerce, which operates certification schemes for cork products.

Assoc. de Empresas de Constru. e Obras Publicas Do Sul
rua Antonia Enes, Nr 9-5, Lisboa 1000
Tel: +351/53 31 93/4/5; *fax:* 351/1 53 28 16

There is an account of construction product certification arrangements in *Building Technical File 30* (July 1990).

CIRIA Special Publication 67 (1989), *The Iberian construction industry: A guide for UK professionals*, describes the background, market, legal and institutional framework, and the Portuguese construction industry.

15.14 SPAIN

E Spain: Espána   **(EC Member State)**
Population: 40 190 000 (80/km^2): 77% urban
Land area: 504 750 km^2
Capital city: Madrid (3 102 846)
Standards body: AENOR Asociación Española Normalización y Certifiación
Tel: +34 1 410 48 51; *fax:* +34 1 410 49 76
ETA bodies: ICCET Instituto de Ciencias de la Construccion Eduardo Torroja
Tel: +34 1 202 04 40; *fax:* +34 1 202 07 00
UEItc Union Española para la Idoneidad Technica en la Construccion
Tel: +34 1 553 16 00; *fax:* +34 1 553 02 44
UEAtc member: IETCC (as above)
CIB full member: IETCC (as above)

Spanish Embassy: 16th floor, Portland House, Stag Place, London SW1E 5SE, UK. *Tel:* 0171 235 5555
Generalitat de Catalunya, the Catalan government, enjoys considerable regional autonomy.

National language: Spanish; regional language: Catalan
Currency: Peseta (Pta)

Asociación Española Normalización y Certificación (AENOR)
Plaza San Juan de la Cruz s/n, E-28003 Madrid
Tel: +34 1 410 48 51; *fax:* +34 1 410 49 76
Spanish national standards body, member of ISO and CEN: a non-governmental body under the tutelage of Industry Ministry, replaced in 1985 former standards institute (IRANOR), which came under CSIC Colegio Superior de Investigación Cientifica (Spanish scientific research council).
Most Spanish standards – UNE (Una Norma Española) – cover product specifications and testing methods, a few engineering design codes. AENOR operates a conformity 'N' mark scheme, but few construction products carry it at present; a standards catalogue is published in Spanish.

Instituto de Ciencias de la Construccion Eduardo Torroja (IETCC)
Serrano Galvache s/n, Apartado 19002, Costillare-Charmartin, E-28033 Madrid
Tel: +34 1 202 04 40; *fax:* +34 1 202 07 00
UEAtc member; one of two designated Spanish technical approval bodies under the EU CPD; full member CIB. An independent institute for building and construction of CSIC, the autonomous council for science and technology under the Ministry of Research and Universities; merged in 1949 with a more specialist cement and concrete institute founded in 1934, taking its name from its first director Professor Eduardo Torroja. Under a 1964 Decree IETcc is empowered to award Agreement certificates – DITs (*documentas de ioneidad technica*); operates two quality mark schemes: CIETSID (steel reinforcement

bars) and CIETAN (precast prestressed concrete beams) for conformance to standards and manufacturer's quality management.

Union Española para la Idoneidad Technica en la Construccion (UEItc)
Plaza San Juan de la Cruz s/n, E-28003 Madrid
Tel: +34 1 553 16 00; *fax:* +34 1 553 02 44
The second Spanish government-designated European technical approval bodies; is associated with AENOR.

Ministerio de Industria, Commercio y Turisomo
Paseo de la Castellana 160, E-28046 Madrid
Tel: +34 1 349 43 98; *fax:* +34 1 457 80 66
One of two Spanish ministries responsible for implementation of the EU Construction Products Directive; has responsibilities for promoting industrial and technological innovations.

Ministerio de Obras Publicas y Transportes (MOPT)
Paseo de la Castellana 67, E-28071 Madrid
Tel: +34 1 553 16 00; *fax:* +34 1 535 23 88
One of two Spanish ministries responsible for implementation of the EU Construction Products Directive; has responsibilities for public buildings and highways, and Spanish construction industry.

Centro de Estudios y Experimentación de Obras Publicas (CEDEX)
Ministerio de Obras Publicas y Urbanismo
Alfonso XII. 3, E-28014 Madrid
Tel: +34/1-467 3708
Spanish public works ministry's research and experimental centre.

Instituto Nacional para la Calidad de la Edificación (INCE)
Plaza de San Juan de la Cruz, Madrid 3
Tel: +34/1-253 3955
National Institute for Quality in Building, attached to Ministerio de Obras Publicas y del Urbanismo.

Laboratorio de Ingenieros del Ejercito
Serrano Jover 2, Madrid 8
Tel: +34/1-248 4800
Testing and standards laboratory of the military engineers.

Centro Nacional de Investigaciones Metalurgicas
Avda. Gregorio del Amo 8, E-28040 Madrid
Tel: +34/1-253 8900
Accredited testing laboratory specializing in metals and metal products, including steel reinforcement.

Colegio General de Colegios Officiales de Ingenieros Industriales
General Arrando 38, Madrid
Tel: +34/1-419 7428
National College of Spanish electrical and mechanical engineers; a Spanish 'college' approximates to a professional institution.

Colegio Nacional de Ingenieros de Caminos, Canakes y Puertos
Almagro 42, Madrid
Tel: +34/1-419 9600
National College of Spanish Civil Engineers.

Consejo General de Colegios Officiales de Aparejadores y Arquitectos Técnicos de España
Paseo de la Castellana 155-la, E-28046 Madrid
Tel: +34 1 570 15 35; *fax:* +34 1 571 28 42
Central organization of Spanish Colleges of 'technical' architects, roughly approximating in UK to building surveyors.

Consejo Superior de los Colegios de Arquitectos
Paseo de la Castellana 12, Madrid
Tel: +34 1-435 2200
Higher council of Spanish Colleges of Architects.

Confederción Nacional de la Construcción
Fiego de Leon 50 - 20, E-28006 Madrid
Tel: +34 1-261 9715
Spanish contractors' federation; at same address ANPCE Asociación de Promotores Constructores de Edifilcios and Asociación Nacional de Constratistas de Obras Publicas (public works contractors).

Centro Informativo de la Construccion (CIC)
Roger de Lluria 117, E-08037 Barcelona
Tel: +34 93 487 04 55; *fax:* +34 93 215 84 15
Regional information centre for Spanish construction industry. Publications include a five-volume collection of standards, codes, and catalogue of building products.

Colegio Oficial d'Aparelladors i Arquitectes Technics de Barcelona
Bon Pastor 5, E-08021 Barcelona
Tel: +34 93 209 82 99; *fax:* +34 93 414 34 34
Catalan college of technical architects; operates Joseph Renart documentation centre.

Institut de Tecnologia de la Construcció de Catalunya (ITECC)
Wellington 19, E-08018 Barcelona
Tel: +34 93-309 34 04; *fax:* +34 93-300 48 52
Principal organization concerned with construction technology in Catalonia.
   There is an account of construction product certification arrangements in *Building Technical File 28* (January 1990). CIRIA Special Publication 67 (1989), *The Iberian construction industry: A guide for UK professionals*, describes the background, construction market, legal and institutional framework, and the Spanish contracting industry.

15.15   SWEDEN

S Sweden: Konungariket Sverige   **(EFTA Member)**
Population: 8 602 000 (19/km^2); 84% urban
Land area: 449 964 km^2
Capital city: Stockholm (672 187)
Standards body: SIS Standardieringskommissionen i Sverige
Tel: +46 8 613 52 00; *fax:* +46 8 11 70 35
UEAtc member: Boverket
Tel: +46 455 53000; *fax:* +46 455 53100
CIB full members: BFR Statens Råd för Byggnadsforskning
Tel: +46 8 617 73 00; *fax:* +46 8 53 74 62
BYGGDOK
Tel: +46 8 34 01 70; *fax:* +46 8 32 48 59
SIB Statens institut för byggnadsforskning
SBJ Svensk Byggjänst
Tel: +46 08 734 50 00; *fax:* +46 08 734 50 99
UICB member: SBJ (as above)

Swedish Embassy: 11 Montague Place, London W1H 2AL, UK
Tel: 0171 917 6400

National language: Swedish
Currency: Swedish krone (SKr)
Sveiges Statskalender, Allmänna Förlaget, Stockholm

Standardieringskommissionen i Sverige (SIS)
PO Box 3295, S-103 66 Stockholm
Tel: +46 8 613 52 00; *fax:* +46 8 11 70 35
National standards body; member of ISO and CEN. In Sweden standards-making mostly done by sectoral bodies including the building standards body: BSI Byggstandardiseringen, Drotting Kristinasväg 73, S-114 28 Stockholm (*Tel:* +46 8 23 72 50; *fax:* +46 8 20 89 93).

Boverket
Drottinggatan 18, PO Box 534, S-37123 Karlskrona
Tel: +46 455 53000; *fax:* +46 455 53100
National Board of Housing, Building and Planning, the government agency for housing provision and finance, building codes and regulation, approval and production control of construction products, and physical planning; UEAtc member.

Statens provningsanstalt (SPA)
(headquarters)
PO Box 857, S-501 15 Borås
Tel: +46 33 16 50 00; *fax:* +46 33 13 55 02
PO Box 56 08, Drottning Kritsinasvägen, S-114 86 Stockholm
Tel: +46/8-10 32 40
Swedish national testing establishment.

Byggnadsstyrrelsen (KBS)
Harlavägen 100, S-106 43 Stockholm
Tel: +46 8 783 10 00; *fax:* +46 8 783 11 80
National Board of Public Building.

National Board of Trade
PO Box 12 09, Birger Jaris torg 5, S-111 82 Stockholm
Tel: +46 8 7 91 05 34; *fax:* +46 8 20 93 24

Statens Energiverk
Liljeholmsvägen 30, S-117 87 Stockholm
Tel: +46 8 744 95 00
National Energy Administration.

Statens Geotekniska Institut
Rosalagsvagen, Stockholm
Tel: +46/8-15 82 74
Headquarters, Linköping
Tel: +46/13-11 51 00

Statens Industriverk
Liljeholmsvägen 30, S-117 86 Stockholm
Tel: +46/8-744 90 00

Statens Kulturråd
PO Box 7843, Långa Raden 4, S-103 98 Stockholm
Tel: +46 8 24 72 60

Statens Naturvårdsverk
PO Box 1302, Smidesvägen 5, S-171 25 Solna
Tel: +46 8 799 10 00

Laboratoriet för Miljökontrol, Uppsala
Tel: +46 18 14 62 80

Statens Vä-Nåmnd
Västerbroplan, Box 12535, S-102 29 Stokcholm
Tel: +46/8-737 58 00

Statiska Centralbyrån (SCB)
Karlvagen100, S-115 81 Stockholm
Tel: +46/8-783 40 00

AB Svensk Byggtjänst (SBJ)
S-171 88 Solna
Tel: +46 08 734 50 00; *fax:* +46 08 734 50 99
Swedish building centre; publishes Swedish building catalogue, AMA (general specification of material and workmanship), and provides a construction book-shop and publications service.

Byggdok
Halsingsgatan 49, S-113 31 Stockholm
Tel: +46 8 34 01 70; *fax:* +46 8 32 48 59

Swedish Institute for Building Documentation; operates an information and documentation service.

Byggindustrin
Byggforlaget, PO Box 5456, Narvavägen 19, S-114 60 Stockholm
Tel: +46/8 663 51 00, *fax:* 46/8 5667 72 78
Principal Swedish construction industry journal.

Chalmers Tekniska Högskola (CTH)
Sektionen för arkitektur industriplanering, S-142 96 Göteborg
School of Architecture, Industrial Architecture and Planning, Chalmers University of Technology.

Kungl Tekniska Högskola (KTH)
Institution för Byggnadstenik, Brinellvägen 34, S-100 44 Stockholm
Tel: +46 8 878 77000

Sveriges Lantbruksuniversitet (SLU)
Institutionen för lantbrukets byggnadsteknik, PO Box 624, S-220
Lund Farm buildings institute, Swedish University for Agriculture.

AB Bostadsgaranti
PO Box 26209, Englebrektsgatan 5, S-100 41 Stockholm
Tel: +46 8 14 20 20; *fax:* +46 8 21 97 83

HSB
PO Box 8310, S-104 20 Stockholm
Tel: +46 8 785 33 87; *fax:* +46 8 785 31 97

Kontrollrädet för betongväror (KRB)
Kronobergsgatan 39, S-112 33 Stockholm
Tel: +46 8 531 780

Stålbyggnadsinstitutet (SBI)
Drotting Kristinas vägen 48, S-114 28 Stockholm
Swedish Institute of Steel Construction.

Svenska Träskyddsinstitutet
PO Box 1507, Stockholm
Tel: +46 8 22 25 40
Swedish timber preservation institute.

Sveriges tegelindustriförening (STIF)
PO Box 5873, S-102 48 Stockholm
Swedish Brick and Tile Industry Association.

Swedish Association of Consulting Engineers
PO Box 222076, S-104 22 Stockholm
Tel: +46 8 654 0860; *fax:* +46 8 650 2972

Swedish Association of Local Authorities
Hornsgatan 15, S-104 20 Stockholm
Tel: +46 8 785 33 87; *fax:* +46 8 785 31 97

Swedish Construction Federation
PO Box 273 08, S-102 54 Stockholm
Tel: +46/8 665 35 00; *fax:* +46/8 662 97 00

Swedish Federation of Housing Cooperatives (SBC)
Svartbäcksgatan 8, S-753 41 Uppsala
Tel: +46 18 69 66 60; *fax:* +46 18 12 34 93

Statens Mät- o. Provråd
Nordtest Examinationscenter Göte Åkerman, Drotting Kristinasvägen
37, Stockholm
Tel: +46 8-11 30 92

15.16 SWITZERLAND

**CH Switzerland: Schweizerische Eidgenossenschaft
– Confédération Suisse – Confederazione
Svizzera**   **(EFTA Member)**
Population: 6 737 000 (163/km^2); 61% urban
Land area: 41 293 km^2
Capital city: Bern (134 398)
Standards body: SNV Schweizerische Normen-Vereingung
Tel: +41 1 384 47 47; *fax:* +41 1 384 47 74
UEAtc member: none
CIB full member: ETH Eidgenössiche Technische Hochschule
Tel: +41 1 377 27 09; *fax:* +41 1 371 80 171
EPFL Ecole Polytechnique Fédérale de Lausanne, Départment
d'Architechture

Swiss Embassy: 16–18 Montague Place, London W1H 2BQ, UK
Tel: 0171 723 0701

Confederation of 22 cantons, each with responsibilities in building
and housing: Zürich; Bern; Luzern; Uri; Schwys; Unterwalden;
Glarus; Zug; Fribourg; Solothurn; Basel (Basel-Stadt and Basel-
Land); Schaffhausen; Appenzell; St Gallen; Graubünden; Aargau;
Thurgau; Ticino; Vaud; Valais; Neuchâtel; Geneva

National languages: French, German, Italian and Romansch
Currency: Swiss franc (Sfr)

Schweizerische Normen-Vereingung (SNV)
Kirchenwet 4, CH-8032 Zürich
Tel: +41 1 384 47 47; *fax:* +41 1 384 47 74
Association for Standardization: member of ISO and CEN. Up to 1978, Swiss
standards carried the initials SNV, since that date SN (Schweizer Norm).

Schweizerischer Ingenieur- und Arckitekten Verein (SIA)
Postfach CH-8039
Tel: +41 1 201 15 70

Society of Swiss Engineers and Architects: principal professional body and main source of professional and technical codes (*normen*) for building and civil engineering; publishes a list of current codes.

Schweizer Baudokumentation (BauDoc)
CH-4223 Blauen
Tel: +41 61 761 41 41; *fax:* +41 761 22 33
Swiss Building Documentation Service for construction and construction products, manages database and building catalogue; has recently developed BAUDOC-DISC for the management of a network of different databases; issues, in French and German, a monthly bulletin *BauDoc Bulletin*; and maintains a mail-order publications service.

Schweizerische Zentralstelle für Baurationalisierung (CRB)
Zemstralstrasse 163, CH-8003, Zürich
Tel: +41 1 451 22 88; *fax:* +41 1 451 15 21
Swiss Centre for Rationalization in Building and Civil Engineering; established by SIA. BSA Federation of Swiss Architects, and SBV Swiss Building Contractors Association; responsible for building cost analyses and standard building description; publishes list of publications.

Swiss Federal Laboratories for Materials Testing and Research (EMPA)
An der Oberlandstr, 129, CH-8600 Dübendorf
Tel: +41 1 823 55 11; *fax:* +41 1 821 62 44
One of Europe's leading testing establishments.

Eidgenössiche Technische Hochschule (ETH)
Institute for Building Materials, CH-8093 Zürich
Tel: +41 1 377 27 09; *fax:* +41 1 371 80 17

Ecole Polytechnique Fédérale de Lausanne (EPFL)
Départment d'Architecture, Institut technique du bâtiment, CH-1001 Lausanne
Research units in the two Swiss universities of technology.

Schweizerische Bauleiter-Organisation (SBO)
c/o SKO, Postfach 383, CH-8042 Zurich (*Tel:* +41 1 361 97 08; *fax:* +41 1 363 16 03)

Schweizerische Zentralstelle für Eigenheim- Wohnbauförderung (SZEM)
Stampfenbachstrasse 69, CH-8035 Zürich
Tel: +41 1 363 22 401
BRE Current Paper 21/70: Evelyn Cibula *Building control in Switzerland* is still a useful source of information on the legal and administrative background to construction in Switzerland with information on procedures and enforcement, the technical content and use of codes and standards issued by SIA.

15.17   UNITED KINGDOM

United Kingdom of Great Britain and
Northern Ireland   **(EC Member State)**
Population: 57 380 000 (235/km^2); 93% urban
Land area: 2 441 100 km^2
Capital city: London (6 575 000)
Standards body: BSI British Standards Institution
Tel: 0171-629 9000; *fax:* 0171-629 0506
ETA body: BBA British Board of Agrément
Tel: 01923 670844; *fax:* 01923 662133
UEAtc: BBA (as above)
Full CIB Members: BRE Building Research Establishment,
Tel: 01923 894040; *fax:* 01923 664099
(for international calls to UK +44)

Principal construction ministry:
DOE Department of the Environment
2 Marsham Street, London SW1
Tel: 0171 276 3000; *fax:* 0171 276 3826

England and Wales: population 49 200 000
Other cities (England): Birmingham (998 000); Leeds (709 000);
Sheffield (532 000); Liverpool (476 000); Manchester (450 000);
Bradford (462 500)
Wales: Cardiff (281 500)
Scotland: Population 5 130 000
Edinburgh (438 000): Glasgow (715 600)
Northern Ireland: population 1 500 000
Belfast (303 800)

National languages: English; Wales: English and Welsh
Currency: Pound sterling (£)

For information on government departments and other public bodies,
see the annual *Whitaker's Almanack*; for UK national associations,
institutes, societies and similar bodies, see the annual *Directory of
British Associations and Associations in Ireland*, CBD Research Ltd,
15 Wickham Rd, Beckenham BR3 2ES (*Tel:* 0181 650 7745; *fax:* 0181
650 0768); for construction industry associations etc, see *CIRIA
Construction Industry Guide* E&FN Spon (1989); for trade associations etc, see P. Millard: *Trade Associations and Professional Bodies
of the UK*, 12th edn (1994).

Many professional, construction industry and manufacturing organizations, directly or through associate bodies or subsidiaries, participate
in standards work, quality assurance and product certification. Their
members may have their factory, laboratory or office management
independently assessed under ISO 9000/EN 29000/BS 5750 by an
accredited independent certification body, or be holders of BS Kitemarks, SGS Yarsley Testguard or other certificate after having had

their products independently tested for conformity to British Standards or BBA Agrément Certificates. They may be accredited as a certification body by NACCB (National Accreditation Council for Certification Bodies) or as a testing laboratory by NAMAS (National Measurement Accreditation Service).

89/106/EEC requires Member States to designate bodies to issue European technical approvals: in UK, DoE has designated BBA; it also designates testing laboratories, inspection bodies and certification bodies as approved bodies under the Directive. DTI (Department of Trade and Industry) has a general responsibility for quality assurance in industry and for testing laboratories through NACCB and NAMAS.

Listed alphabetically under initials in common use are certification, quality assurance, research and similar technical organizations, associated government departments, professional and industry bodies, with where appropriate, notes on activities and responsibilities. *Architect's Reference Annual*, issued by RIBA Publications, and other guides list further government, professional and industry organizations.

Association of British Certification Bodies (ABCM)
C/o LRQA, Norfolk House, Wellesley Road, Croydon CR9 2DT
Tel: 0181 688 6883
Recognized by NACCB as forum and spokesbody for views of its 35 members.

Association of Consulting Architects (ACA)
Buchanan's Wharf, Redcliffe Backs, Bristol BS1 6HT
Tel: 0117 929 3379; *fax:* 0117 925 6008

Association of Consulting Engineers (ACE)
12 Caxton Street, London SW1H 0QL
Tel: 0171 222 6557; *fax:* 0171 222 0750

Architects Registration Council of the United Kingdom (ARCUK)
73 Hallam Street, London W1N 6EE
Tel: 0171 580 5861; *fax:* 0171 436 5269

Association of Quality Management Consultants (AQMC)
4 Beynes Road, Olivers Battery, Winchester SO22 4JW
Tel: 01962 866969; *fax:* 01962 866969
Members are professional quality management consultants to industry and commerce.

Certification Services (ASTA)
23–24 Market Place, Rugby CY21 3DU
Tel: 01788 578435

British Aggregate Construction Materials Industries (BACMI)
156 Buckingham Palace Road, London SW1W 9TE
Tel: 0171 730 8194
Producers of ready-mixed concrete belong to QSRMC scheme.

British Approvals Service for Electric Cables (BASEC)
PO Box 390, Breckland, Linford Wood, Milton Keynes MK14 6LL
Tel: 01908 6911210/691121; *fax:* 01908 315555

British Board of Agrément (BBA)
PO Box 195, Bucknalls Lane, Garston, Watford WD2 7NG
Tel: 01923 670844; *fax:* 01923 662133
Designated single UK technical approval body under the EC CPD, and UK spokesbody on EOTA; member of UEAtc. BAA issues quarterly its *Index of Current BBA Publications*, copies being obtained from BBA Publications Department, PO Box 195, Bucknalls Lane, Garston, Watford WD2 7NG (*Tel:* 01923 670844; *fax:* 01923 662133).

Building Centre (BC)
2 Store Street, London WC1E 7BT
Tel: 0171 637 1022; *fax:* 0171 580 9641; Information Service: 0344 884999; Building Bookshop: 0171 637 3151; *fax:* 0171 636 3628

British Cement Association (BCA)
Century House, Telford Avenue, Crowthorne RG11 6YS
Tel: 01344 762676; *fax:* 01344 761214
Formerly Cement and Concrete Association; recently restructured as The Centre for Concrete Training.

CERAM Research (British Ceramic Research Limited) (BCRA)
Queens Road, Penkhull, Stoke-on-Trent ST4 7LQ
Tel: 01782 45431; *fax:* 01782 412331
Category O NAMAS-accredited testing laboratory for ceramic and other construction products. Not an approved certification body; see CICS.

Brick Development Association (BDA)
Woodside House, Winkfield, Windsor SL4 2DX
Tel: 01344 885651: fax 01344 890129
Serves brickmakers and users; technical publications; not a certification body; brickmaker members may be assessed through CICS.

British Electro-technical Approvals Board (BEAB)
March House, The Green, 9/11 Hersham, Walton-on-Thames KT12 5NA
Tel: 019322 44401
Operates Safety Mark Schemes for 'white goods' and 'brown goods' with BSI QA.

British Employers' Confederation (BEC)
82 New Cavendish Street, London W1M 8AD
Tel: 0171 580 5588
Represents contractors in standards-making and other construction industry activities directly and through associate organizations including BWF and HBF.

(National Council of) Building Material Producers (BMP)
26 Store Street, London WC1E 7BT
Tel: 0171 323 3770; *fax:* 0171 323 0307

Representative body for building materials producers at national level; at European level directly and through CEPMC (Council of European Producers of Materials for Construction), of which a founder member.

British Non-Ferrous Metals Federation (BNFMF)
10 Greenfield Crescent, Birmingham B15 3AU
Tel: 0121 456 3322
Technical Centre, Grove Laboratory, Detchworth Road, Wantage OX12 9BJ
Tel: 01235 72992

British Precast Concrete Federation (BPCF)
60 Charles Street, Leicester LE1 1FB
Tel: 0116 253 6161; *fax:* 0116 251 4568
Members include ACBA (Aggregate Concrete Block Association); CPA (Concrete Pipe Association); PCFA (Precast Concrete Frame Association); PCCA (Precast Concrete Cladding Association); and Interpave (Concrete Block Paving Association).

British Plastics Federation (BPF)
6 Bath Place, Rivington Street, London EC2A 3JE
Tel: 0171 457 5000; *fax:* 0171 457 5045

Building Research Establishment (BRE)
Garston, Watford WD2 7RJ
Main switchboard 01923 894040; *fax:* 01923 664010
Technical Consultancy & Advisory Service
Tel: 01923 664800
Publications Sales
Tel: 01923 664444
Scottish Laboratory, Kelvin Road, East Kilbride, Glasgow G75 0RZ
Tel: 0135 52 33001; *fax:* 0135 52 41895
Since April 1990 an Executive Agency of DoE (Department of the Environment), BRE is the principal UK organization undertaking research into building and construction and control of fire. BRE is a NAMAS Category O accredited testing laboratory.

The annual *BRE Bookshop Catalogue* lists BRE information products: books and reports; information papers on recent research results; digests of essential reference material; good building guilds and audiovisual material. BRE Bookshop, Watford WD2 7JR (*Tel:* 01923 664444; *fax:* 01923 664400).

British Standards Institution (BSI)
399 Chiswick High Road, London W4 4AL
BSI Head Office and BSI Standards
(*Tel:* 0181 996 9000; *fax:* 0181 996 7400),
also BSI Sales & Customer Services
(*Tel:* 0181 996 7000; *fax:* 0181 996 7001),
BSI Membership Service
(*Tel:* 0181 996 7002; *fax:* 0181 996 7001),
BSI Information Services, including Technical Help to Exporters
(*Tel:* 0181 996 7111; *fax:* 0181 966 7048),

BSI Library Services
(*Tel:* 0181 996 7004; *fax:* 0181 996 7048),
BSI Translations & Language Services
(*Tel:* 0181 996 7222; *fax:* 0181 996 7047),
BSI Copyright & Conferences
(*Tel:* 0181 996 7070; *fax:* 0181 996 7048),
BSI Electronic Products Help Desk
(*Tel:* 0181 996 7333; *fax:* 0181 996 7047),
PO Box 375, Milton Keynes MK14 677
BSI Quality Assurance
(*Tel:* 01908 220908; *fax:* 01908 220671),
BSI Product Certification
(*Tel:* 01908 312636; *fax:* 01908 695157),
Maylands Avenue, Hemel Hempstead HP2 4SQ
BSI Testing
(*Tel:* 01442 230442; *fax:* 01442 231442)
UK national standards body; members of ISO and CEN. British Standards are updated in a monthly insert in *BSI News*. During summer 1994, BSI moved its headquarters, and its standards-making and a number of other activities, to new, specially equipped facilities at Chiswick in West London near Gunnersbury station on the London District Underground railway Richmond line. BSI Quality Assurance and Products Certification remains at Milton Keynes, and BSI Testing at Hemel Hempstead.

Building Services Research and Information Association (BSRIA)
Old Bracknell Lane, Bracknell RG12 7AH
Tel: 01344 426511; *fax:* 01344 487575
NAMAS Category O and Category I accredited testing laboratory; serves building services industry and professions; operates Eurocentre for information and publication service on codes, regulations and standards relating to the Single European Market including translations on VDI (German Association of Engineers) codes.

Bureau Veritas Quality International (BVQI)
70 Borough High Street, London SE1 1XF
Tel: 0171 378 8113; *fax:* 0171 378 8014
Certification body accredited by NCCAB and NL RvC.

British Wood Preserving and Damp-proofing Association (BWPDA)
Building 6, The Office Village, 4 Romford Road, Stratford, London E15 4EA
Tel: 0181 519 2588; *fax:* 0181 519 3444
Operates with BSI QA Registered Firm schemes for fire retardant of timber etc.

Cement Admixtures Association (CAA)
Harcourt, The Common, Kings Langley WD4 8BL
Tel: 01923 264314; *fax:* 01923 270778
Member of EFCA (European Federation of Concrete Admixtures Association).

UK Certification Authority for Reinforcing Steels (CARES)
Oak House, Tubs Hill, Sevenoaks TN13 1BL
Tel: 01732 450000; *fax:* 01732 455917
NACCB-accredited certification body for reinforcing steels, including deformed bars, covering all stages from manufacture to delivery on site.

Cathedral Fabric Commission
83 London Wall, London EC2H 5AA
Tel: 0171 638 0971; *fax:* 0171 638 0184

Copper Development Association (CDA)
Orchard House, Mutton Lane, Potters Bar EN6 3AP
Tel: 01707 50711; *fax:* 01707 42769

Centre for Window and Cladding Technology
University of Bath, Claverton Down, Bath BA2 7AY
Tel: 01225 826541; *fax:* 01225 826556

Certifire Ltd
Fire Prevention Certification, 101 Marshgate Lane, London E15 2NQ
Tel: 0181 555 3234; *fax:* 0181 519 3029

Chartered Institution of Building Services Engineers (CIBSE)
Delta House, 222 Balham High Street, London WS12 9BS
Tel: 0181 675 5211; *fax:* 0181 675 5449
Takes major initiative in development of quality assurance for building services sector, for which it is a major source of technical information through *CIBSE Guides* etc.; responsible for National Engineering Specification.

Construction Industry Council (CIC)
26 Store Street, London WC1E 7BT
Tel: 0171 637 8692

Ceramic Industry Certification Scheme Ltd (CICS)
Queens Road, Penkhull, Stoke-on-Trent ST4 74Q
Tel: 01782 45431
NACCB-accredited certification body for ceramic industry.

Construction Industry Information Group (CIIG)
26 Store Street, London WC1E 7BT
Tel: 0171 465 3253

Chartered Institute of Building (CIOB)
Englemere, Kings Ride, Ascot SL5 8BJ
Tel: 01344 23355; *fax:* 01344 23467
Takes major initiative in development of quality assurance for building construction and management, for which it is a major source of technical and management information; operates an Information Resource Centre and bookshop.

Construction Industry Research and Information Association (CIRIA)
6 Storey's Gate, London SW1P 3AU
Tel: 0171 222 8891; *fax:* 0171 222 1708

Independent private-sector research and information organization whose objective is 'to improve the performance of all concerned with construction by prompting the exchange of experience and collaborative, and by producing and disseminating information'. CIRIA has given a lead to the construction professions and industry in ensuring the smooth and effective implementation of QA in construction. Publications include *CIRIA News*, in which the Association's research and information activities are reported, and a wide range of research and technical reports in *CIRIA Publications Catalogue*.

Construction Industry Standing Conference (CISC)
26 Store Street, London WC1E 7BT
Tel: 0171 323 5270; *fax:* 0171 580 9641

Council for Registered Gas Installers (CORGI)
4 Elmwood, Chineham Business Park, Crockford Lane, Basingstoke RG24 OWG
Tel: 01256 708135; *fax:* 01256 708144
Under Gas Safety (Installation and Use) (Amendment) Regulations 1990, all employers and self-employed individuals installing, maintaining, repairing, surveying or specifying piped gas fittings must be CORGI registered.

Construction Quality Assurance Ltd (CQA)
Arcade Chambers, The Arcade, Market Place, Newark NG24 1UD
Tel: 01636 708700; *fax:* 01636 708766

Department for Education
Sanctuary Buildings, Great Smith Street, London WC1P 3BT
Tel: 0171 925 5000

Department of the Environment (DoE)
2 Marsham Street, London SW1P 3EB
Tel: 0171 276 3000; *fax:* 0171 3826; *DoE Euronews Construction*: 0171 276 6597
Responsible for sponsorship of the construction, building materials and aggregates industries, including government policies on standards, quality assurance, and agrément, and international matters relating to construction; also for policy, preparation and amendment of Building Regulations in England and Wales. DoE issues a monthly *Construction Monitor*, which incorporates *Euronews Construction* and gives up-to-date information on matters relating to quality, the environment and research. Relevant construction statistics are also published in *Construction Monitor*.

Department of Health (DoH)
79 Whitehall, London SW1A 2NS
Tel: 0171 210 5983

Det norske Veritas Quality Assurance Ltd (DNV)
112 Station Road, Sidcup DA15 7BU
Tel: 0181 309 7477; *fax:* 0181 309 5907
NACCB-accredited certification body.

Department of Trade and Industry (DTI)
Kingsgate House, 66–74 Victoria Street, London SW1E 6SW
Tel: 0171 215 5000; Single European Market: 0181 200 1992

Responsible for Government policy on standards, certification, quality assurance and testing with particular reference to technical barriers to trade and EEC Directives Publishes *The Register of Quality Assured United Kingdom Companies.*

Electrical Equipment Certification Service (EECS)
Harpur Hill, Buxton SK17 9JN
Tel: 01298 26211

Electricity Association Quality Assurance (EAQA)
30 Millbank, London SW1P 4RD
Tel: 0171 828 9227 or 0171 834 2333; *fax:* 0171 828 9237
Specializes in field of electricity power generation, transmission and distribution; closely associated with Electricity Association, industry's trade association.
    NAMAS Category O accredited testing laboratory; operates EECS (Electrical Equipment Certification Scheme).

English Heritage (EH)
26 Savile Row, London W1X 2BT
Tel: 0171 734 6010
Also Historic Buildings and Ancient Monuments Commission for England.

Engineering Contractors' Quality Assurance Ltd (EQUAL)
ECSA House, 34 Palace Court, London W2 4HY
Tel: 0171 229 1266

Furniture Industry Research Association (FIRA)
Maxwell Road, Stevenage SG1 2EW
Tel: 01438 727607
Serves furniture industry; NAMAS Category O accredited testing laboratory.

Federation of Master Builders (FMB)
14–15 Great James Street, London WC1 3DP
Tel: 0171 242 7583; *fax:* 0171 404 0296

Fire Protection Association (FPA)
140 Aldersgate Street, London EC1A 4HX
Tel: 0171 606 3757; *fax:* 0171 600 1487

FRS
See BRE (Building Research Establishment).

Heating, Ventilating and Air Conditioning Manufacturers' Association (HEVAC)
Nicholson House, High Street, Maidenhead SL6 1LF
Tel: 01628 34667/8

Her Majesty's Stationery Office (HMSO)
HMSO Books, 51 Nine Elms Lane, London SW8 5DK
Tel: 0171 873 2000; *fax:* 0171 873 8463
Publisher and sales office of official international, European and UK publications and journals.

Her Majesty's Treasury (HMT)
Parliament Square, London SW1P 3AG
Tel: 0171 270 5366
UK government department with responsibilities for implementation of EU public procurement Directives.

Home Office (HO)
50 Queen Anne's Gate, London SW1 9AT
Tel: 0171 273 3000; information services: 0171 273 3043; fire department: 0171 273 3406

Health & Safety Executive (HSE)
Barnards House, 1 Chepstow Place, London W2 4TF
Tel: 0171 243 6000
UK executive agency of HSC (Health and Safety Commission) with statutory responsibilities for health and safety at work, including construction and for implementation of relevant EU Directives.

Institution of Civil Engineers (ICE)
1–7 Great George Street, London SW1P 3AA
Tel: 0171 222 7722; *fax:* 0171 222 7500

Institution of Electrical Engineers (IEE)
Savoy Place, London WC2R 0BL
Tel: 0171 240 1871; *fax:* 0171 240 7735

Institution of Mechanical Engineers (IMechE)
1 Birdcage Walk, London SW1H 9JJ
Tel: 0171 222 7899; *fax:* 0171 222 4557

Institution of Structural Engineers (IStructE)
11 Upper Belgrave Street, London SW1X 8BH
Tel: 0171 235 4535; *fax:* 0171 235 4294

Local Authorities Coordinating Body on Food and Trading Standards (LACOTS)
PO Box 6, 1A Robert Street, Croydon CR9 1LG
Tel: 0181 688 1996; *fax:* 0171 680 1509
Member authorities have responsibilities for enforcement of a number of statutory regulations implementing EU Directives.

Loss Prevention Council (LPC)
140 Aldersgate Street, London EC1A 4HY
Tel: 0171 606 1050; *fax:* 0171 600 1457
Representative organization of Association of British Insurance and Lloyd's; issues annual list of approved fire protection and safety and security products, list of security system installers approved by the National Approval Council for Security Systems, and publications of the Fire Protection Association.

Loss Prevention Council Technical Centre (formerly FIETO)
Melrose Avenue, Borehamwood WD6 2BJ
Tel: 0181 207 2345; *fax:* 0181 207 6305

NAMAS-accredited Category O fire testing laboratory; also Category O testing laboratory in fields of corrosion, dimensional and environmental testing, fire, mechanical and performance testing, physical and safety testing; NACCB-accredited certification body.

Lloyd's Register Quality Assurance Ltd (LRQA)
Norfolk House, Wellesley Road, Croydon CR9 2DT
Tel: 0181 688 6882; *fax:* 0181 681 8146

NACCB accredited certification body operating sector-based industrial and single-firm certification schemes for quality management systems, project management, product conformity certification, stockist quality systems and quality assurance personnel training; also accredited in the Netherlands by RvC (Raad voor Certificatie); a subsidiary company of Lloyd's Register of Shipping.

Mastic Asphalt Council & Employers' Federation (MACEF)
Technical advisory centre, 6–8 The Broadway, Bexleyheath DA6 7LE
Tel: 0181 299 0414; *fax:* 0181 298 0381

Mortar Producers' Association (MPA)
74 Holly Walk, Leamington Spa CV32 4JD
Tel: 01916 38611

National Accreditation Council for Certification Bodies (NACCB)
13 Palace Street, London SW1E 5HS
Tel: 0171 233 7111; *fax:* 0171 233 5115

Set up in 1985 by the Department of Trade and Industry to be responsible for the assessment of certification bodies applying for Government accreditation with responsibilities wider than construction. The council carries out independent audits on these bodies to ensure that they are conforming consistently to the appropriate standard. NACCB participates actively in the European Accreditation of Certification (EAC), whose aim is the setting-up of a single European accreditation system.

National Measurement Accreditation Service Executive (NAMAS)
National Physical Laboratory, Teddington TW1 0LW
Tel: 0181 943 6311; *fax:* 0181 943 7134

Combines the operations of BCS (British Calibration Service) and NATLAS (National Testing Laboratory Accreditation Scheme), on behalf of DTI (Department of Trade Industry). Laboratories are listed in the *NAMAS Directory*.

National Housing and Town Planning Council
14–18 Old Street, London EC1V
Tel: 0171 251 2363; *fax:* 0171 608 2830

National House-Building Council (NHBC)
Buildmark House, Chiltern Avenue, Amersham HP6 5AP
Tel: 01494 434477; *fax:* 01494 728521

Independent council setting technical standards for housebuilding additional to health and safety requirements of Building Regulations; accepted by central government as an alternative agency to district councils for technical control under building regulations.

National Inspection Council for Electrical Installations Contracting (NICEIC)
36/37 Albert Embankment, London SE1 7U
Tel: 0171 582 7746
Maintains register of approved electrical installation contractors.

Northern Ireland Office (NIO)
Stormont, Belfast BT4 3SS
Tel: 01232 520700

National Joint Consultative Committee for Building (NJCC)
18 Mansfield Street, London W1M 9FG
Tel: 0171 580 5588

National Physical Laboratory, Division of Materials Metrology (NPL)
Teddington TW11 0LW
Tel: 0181 977 3222
NAMAS Category O accredited laboratory

National Trust for Places of Historic Interest or Natural Beauty (NT)
42 Queen Anne's Gate, London SW1H 9AS
Tel: 0171 222 9251; *fax:* 0171 222 5097

Ordnance Survey (OS)
Romsey Road, Maybush, Southampton SO9 4DH
Tel: 01703 792000

Patent Glazing Contractors' Association (PGCA)
13 Upper High Street, Epsom KT17 4QY
Tel: 01372 729191; *fax:* 01372 729190

Partitioning Industry Association (PIA)
692 Warwick Road, Solihull B91 3DX
Tel: 0121 705 9270; *fax:* 0121 711 2892

Research Association for the Paper and Board, Printing and Packaging Industries (PIRA)
Randalls Road, Leatherhead KT22 7RU
Tel: 01372 376161

Paint Research Association (PRA)
Waldegrave Road, Teddington TW11 8LD
Tel: 0181 977 4427; *fax:* 0181 943 4705
NAMAS Category O accredited testing laboratory.

Plastics and Rubber Institute (PRI)
11 Hobart Place, London SW1W 0HL
Tel: 0171 245 0555; *fax:* 0171 829 1379

Quality Scheme for Ready Mixed Concrete (QSRMC)
3 High Street, Hampton TW12 2SQ
Tel: 0181 941 0273; *fax:* 0181 979 4558
Independent NACCB-accredited specialist certification body.

Rubber and Plastics Research Association (RAPRA)
Shawbury, Shrewsbury SY4 4NR
Tel: 01939 250383; *fax:* 01939 251187
NAMAS Category O accredited testing laboratory for polymeric products.

Royal Institute of British Architects (RIBA)
66 Portland Place, London W1N 4AD
Tel: 0171 580 5533; *fax:* 0171 255 1541

British Architectural Library (BAL)
fax: 0171 631 1802
BAL Drawings Collection
21 Portman Square, London W1H 9HF
Tel: 0171 580 5533
RIBA Publications Ltd, 39 Moreland Street, London EC1Y 8BB
Tel: 0171 251 5885
RIBA Bookshop: 66 Portland Place, London W1N 4AD
Tel: 0171 251 0791

Royal Institution of Chartered Surveyors (RICS)
12 Great George Street, London SW1P 3AD
Tel: 0171 222 7000; *fax:* 0171 222 9430

Royal Town Planning Institute (RTPI)
26 Portland Place, London W1N 4BE
Tel: 0171 636 9107

Steel Construction Institute (SCI)
Silwood Park, Ascot SL5 7QN
Tel: 01344 23345; *fax:* 01344 22944
Independent research and development organization; leading European centre for steel design and construction technology; operates technical advisory, library information and publication services

Steel Construction Quality Assurance Scheme Ltd (SCQAS)
4 Whitehall Court, London SW1A 2ES
Tel: 0171 976 1634
NACCB-accredited specialist third-party quality assurance certification body.

Scottish Office Environment Department
New St Andrews House, St James Centre, Edinburgh EH1 3SZ
Tel: 0171 556 8400
Central government department responsible for construction, housing and planning, including building regulations and roads, in Scotland.

SGS Yarsley International Certification Services Limited
Trowers Way, Redhill RH1 2JM
Tel: 01737 768445; *fax:* 01737 772845
NACCB-accredited certification body; NAMAS-accredited category O testing laboratory.

Society for the Protection of Ancient Buildings (SPAB)
37 Spital Square, London E1 6DY
Tel: 0171 377 1644

Swedish Finnish Timber Council
21 Carolgate, Retford DN22 6BZ
Tel: 01777 706616; *fax:* 01777 704695

Timber Research and Development Association (TRADA)
Stocking Lane, Hughenden Valley, High Wycombe HP14 4ND
Tel: 01494 565484; *fax:* 01494 565487
R&D association supported by the timber and timber products industries;
TRADA Technology Ltd, accredited by NAMAS as a Category O testing
laboratory.

Transport Research Laboratory (TRL)
Old Wokingham Road, Crowthorne, Berkshire RG11 6AU
Tel: 01744 770203; *fax:* 01344 770193

UK Certification Authority for Reinforcing Steels
See CARES.

Warrington Research Centre (WFRC) (hazard evaluation, loss
prevention)
Holmesfield Road, Warrington WA1 2DS
Tel: 01925 55116; *fax:* 01925 55419
NAMAS Category O accredited testing laboratory; with BSI QA operates
Certifire scheme for passive fire products.

London laboratory, 101 Marshgate Lane, London E15 2NQ
fax: 0181 519 3029
NAMAS Category O and Category I accredited testing laboratory.

Water Industry Certification Scheme (WICS)
1 Queen Anne's Gate, London SW1H 9BT
Tel: 0171 222 8111

Welsh Office (WO)
Cathays Park, Cardiff CF1 3NQ
Tel: 01222 825111

Water Research Centre (WRc)
Frankland Road, Blagrove, Swindon SN5 8YF
Tel: 01793 511711; *fax:* 01793 51172
Operates Water Byelaws Approval Scheme.

Water Services Association of England & Wales (WSAEW)
1 Queen Anne's Gate, London SW1H 9BT
Tel: 0171 222 8111; *fax:* 0171 222 12811

YQAF
See SGS Yarsley Quality Assured Firms Ltd.

Zinc Development Association (ZDA)
42 Weymouth Street, London W1N 3LQ
Tel: 0171 499 6636; *fax:* 0171 499 1555

Terms and definitions relating to codes, regulations, standards, certification and testing

16

16.1 ENGLISH

Many of the definitions are extracted from official documents, international or national standards. The following authoritative glossaries give terms in English relating to building and civil engineering, including quality assurance: BS 4778: *Quality vocabulary* Part 2:1991 *Quality concepts and related definitions*; BS 6100: *Glossary of building and civil engineering terms*, especially Part 1, Section 1.7 Performance and characteristics.

acceptance check
Check by responsible body to ensure there is adequate manufacturer's quality control, and current production in conformity with the Agrément specifications. (UEAtc)

acceptance testing
Determination of compliance with performance requirements by simplest reliable means. (BS 6100)

accreditation
Recognition of the competence of a certification or inspection body, or a testing laboratory, according to a legal procedure established by a Member State.

acknowledged rule of technology
Technical provision acknowledged by a majority of representative experts as reflecting the developed stage of technical capability at a given time as regards products, processes and services, based on relevant consolidated findings of science, technology and experience. (The Construction Products Regulations 1991)

agrément
This term, meaning 'approval' or 'acceptance', was the name used initially for the approval system introduced by CSTB Centre Technique

et Scientifique du Bâtiment in France to provide an independent evaluation of new products. It was adopted by the British Board of Agrément, when it was extended to cover products and techniques not yet the subject of a British Standard because of their novelty, applications other than those set out in the relevant BS, or having a performance in one or more respects superior to that required in the Standard.

Agrément Certificate
Certificate issued by BBA, giving an independent evaluation of products and techniques for their fitness for use in specific conditions.

appropriate attestation procedure
The procedure (being one of the procedures set out in paragraphs 2 and 4 of Schedule 3) indicated in relation to a construction product in the relevant technical specification or in the publication of that specification. (The Construction Products Regulations 1991)

approved body
Approved laboratory, a certification body or an inspection body. (The Construction Products Regulations 1991: see also CPD Ch.VII: 89/106/EEC)

approved laboratory
Testing laboratory designated for the purposes of the Directive by a Member State, and notified by that Member State to the European Commission. (The Construction Products Regulations 1991)

attestation
Formal confirmation; verification of the execution of a legal document.

attestation of conformity
CPD Ch.V: 89/106/EEC lays down the procedure for attestation of conformity with requirements of a technical specification by a manufacturer. Attestation is dependent on:

- the manufacturer having a factory production control system to ensure that production conforms with the relevant technical specifications;
- for particular products indicated in the relevant technical specifications, in addition to a factory production control system, an approved certification body being involved in assessment and surveillance of the production control or the product itself.

building control system
System through which 'the design and construction of buildings and the provision of services, fittings and equipment in or in connection with buildings' is regulated for health and safety, and other matters considered by government to be in the public interest.

building guarantee
Warranty by the contractor, backed by insurance cover, a bank guarantee or contract retention, covering rectification of stated defects within a specified period.

building inspector
Person legally authorized to inspect work at design and construction stages for conformity with building and associated legislation. (UN ECE Working Party on Building, 1980)

calibration
Set of operations which establish, under specified conditions, relationship between values indicated by a measuring instrument or measuring system, or values represented by a material measure, and the corresponding known value of a measurand. (*Vocabulaire International de Métrologie*)

CE mark
Mark affixed on the product, a label, its packaging, or accompanying commercial documents (delivery ticket), certifying conformity with the relevant standard or technical approval.

An indication of compliance of the product with the relevant technical specifications, especially the harmonized standards or European technical approvals. (Draft EC SCC Guidance Paper 2, 1990)

certify
To make (a thing) certain; to guarantee as certain; to give certain information of; to declare or attest by a formal or legal certificate; to assure; to vouch for. (*Shorter Oxford English Dictionary*)

certificate of conformity
Document attesting that a product or service complies with a given technical specification.

certificate of conformity EC
Certificate of conformity issued by a certification body in accordance with paragraphs 1 and 3 of Schedule 3, or under the Directive as implemented under the law of a Member State other than the United Kingdom. (The Construction Products Regulations 1991)

certification
The term 'certification' may cover product conformity certification, product approval of specified uses, certification of a supplier's quality management system, or certification of personnel involved in quality verification. (DTI/BSI Memorandum of Agreement)

Act of licensing by a document formally attesting the fulfilment of conditions. (BS 4778:1991 = ISO 8402)

Independent attestation of conformity of a product to a technical specification by an approved certification body. (89/106/EEC Art.18)

certification body
An impartial body, which may be a public agency or a private organization, whose competence and impartiality has been formally recognized by a national authority, responsible to an appropriate government ministry. In the UK, that body is NACCB, the National Accreditation Council for Certification Bodies, in which the interests of all parties concerned with the functioning of the system are represented, the appropriate ministry being DTI, the Department of Trade and Industry. The standards that should be used for assessment are:

- EN 45001, *General criteria for certification bodies operating product certification*; and
- EN 45012, *General criteria for certification bodies operating quality systems certification*.
  A body designated as a certification body for the purposes of the Directive by a Member State, and notified by that Member State to the European Commission. (The Construction Products Regulations 1991)

certification of product conformity
Act of certifying by means of a certificate of conformity that a product is in conformity with specific standards, or technical specifications. (ISO Guide 2, quoted in NACCB Criteria of Competence, undated)

certification of production control
Confirmation that all products are manufactured under a scheme of supervision, inspection and control.

certifying body (agrément)
Body issuing the product certification in support of agrément, generally the agrément institute, but the task in whole or part may be mandated to independent bodies. (UEAtc Rule T.03 1988)

class
A range of levels comprising between a lower level and an upper level; a way of expressing the behaviour of a work or product, under a certain action, by taking into consideration in conventional manner several factors which play a role in this behaviour, and give rise to tests or calculations. (EC SCC Guidance Paper 11: 1991)

classes of performance
Definition of grades in European technical specifications. (89/106/EEC, Ch.II, Art.6.3)

client
The party for whom work is done; as well as successive owners; tenants are included under this definition. (NEDO, *Build Report*, 1988)

code of practice
Document setting down the theory and accepted principles of design and/or construction practices for achieving certain aims. A code of practice is constantly under review as the science and art of construction develops. (RIBA Practice Notes, April 1986)

Rule describing recommended practices for design, manufacturing, setting out, maintenance or use of equipment, installations, structures or products. (UN ECE Working Party on Building, 1980)

conformity mark
A mark signifying that a product satisfies Art.4.2 89/106/EEC. The mark consists of the symbol CE; and has to be accompanied by the name, or mark of the producer and, where appropriate, other details and references.

The EC mark of conformity referred to in regulation 5, consisting of the symbol 'CE' of which a form is shown for the purposes of illustration in Schedule 1. (The Construction Products Regulations 1991)

common technical specifications
Technical specification laid down in accordance with a procedure recognized by the Member States to ensure uniform application in all Member States, which has been published in the *Official Journal of the European Communities*. (89/106/EEC, Annex III)

competent authority
An authority which entitles a developer to proceed with a project.

consensus
General agreement, characterized by the absence of sustained opposition, to substantial issues by any important part of the concerned interests, and by a process which involves seeking to take into account the views of all parties concerned, and to reconcile any conflicting arguments. Note: consensus need not imply unanimity. (BS 0: Part 1: 1991)

construction product
Any product which is produced for incorporation in a permanent manner in construction works, including both buildings and civil engineering works. (89/106/EEC, Art.1(2))

Any product which is produced for incorporation in a permanent manner in works. (The Construction Products Regulations 1991)

contractors' 'all risks' insurance
Material damage policy which protects the contractor and employer against loss and damage to the contract works during erection and until completed. (NEDO, *Build Report*, 1988)

damage
Material manifestation of a defect, not being a malfunction caused by *force majeure*, sabotage, vandalism, erroneous utilization or maintenance. (CIB W87)

decennial insurance
Any form of latent defects protection insurance given on a 10-year forward basis; in France, *dommage ouvrage*, the basic compulsory decennial policy for building owners, and the *responsabilité civile décennale*, taken out by members of the construction team, are examples.

Decision (EC)
An item of Community legislation binding in entirety on those to whom addressed whether Member States, companies or individuals; may impose financial obligations enforceable in national courts. (DTI, *The Single Market: The Facts*, 1992)

declaration
Statement by manufacturer regarding the conformity on the basis detailed in 89/106/EEC Annex III 2 (ii).

declaration of conformity (EC)
Declaration of conformity, issued by a certification body in accordance with paragraphs 4 and 5 of Schedule 3, or under the Directive as implemented under the law of a Member State other than the United Kingdom. (The Construction Products Regulations 1991)

declaration of recognition
Confirmation of certification relating to Agrément. (UEAtc Rule T.03 1988)

deemed-to-satisfy-provision
Provision that a particular requirement will be regarded as being met by a product or service which complies with a given technical specification. (UN ECE Working Party on Building, 1980)

deductible
First amount of claim not covered. (NEDO, *Build Report*, 1988)

defect
Either the building does not conform to 'recognized' rules (codes, norms . . .), or the instructions for maintenance are inadequate or lacking, or both. (CIB W87)

derogation
Partial abrogation, or repeal of a law; term used in European legislation to indicate a dispensation from, or relaxation of, requirements of a Council Directive; from *déroger* [F] 'to depart from the law'.

developer
Conglomerate term embracing designers, other professional consultants, main and subcontractors, and in certain circumstances regulatory authorities; but not including manufacturers and suppliers of building materials and components. (NEDO, *Build Report*, 1988)

Directive (EU)
A set of rules promoted by the Commission, and adopted by the Council of Ministers after appropriate consultation, binding in respects of results to be achieved; but left to individual Member States to implement by an appropriate administrative, or legislative procedure; from *directive* [F] 'rule of conduct; general lines of a policy'. (DTI, *The Single Market: The Facts, 1992*)

(UEAtc have abandoned using the term for a technical document serving as basis for assessment of products.)

Directive, the
Council Directive 89/106/EEC on the approximation of laws, regulations and administrative procedures of the Member States relating to construction products. (The Construction Products Regulations 1991)

durability
The ability of an item to perform its required function under stated conditions of use, and under stated conditions of preventable or corrective maintenance until a limiting state is reached. (BS 5750)

duty of care
Legal duty to take reasonable care, or avoid acts or omissions which are likely to injure those to whom the duty is owed; breach of this duty results in liability for negligence. (NEDO, *Build Report*, 1988)

economically reasonable working life
The period of time during which the performance of the works will be maintained at a level compatible with the fulfilment of the essential requirements. (EC Interpretative Document for the Essential Requirement: Mechanical Resistance and Stability, 1991)

electromagnetic disturbance
Any electromagnetic phenomenon which may degrade the performance of a device, unit of equipment or system. An electromagnetic disturbance may be electromagnetic noise, an unwanted signal or a change in the propagation medium itself. (Electromagnetic Compatability Directive 89/336/EEC)

enabling powers
Powers given by Statute to a government department, or other body, enabling it to undertake certain measures, e.g. making regulations. (Legal)

enforcement authority
The Secretary of State, any other Minister of the Crown in charge of a Government department, any such department or Northern Ireland department, and any authority or council on whom functions under these Regulations are conferred by regulation 15. (The Construction Products Regulations 1991)

error or omission
Any departure from correct building practice, any absence of adequate maintenance instructions. (CIB W87)

essential requirements
Requirements regarding safety, health and certain other aspects in the general interest that the construction works can meet: Annex III; six essential requirements are set out in Annex I. (89/106/EEC)

Requirements applicable to works which may influence the technical characteristics of a construction product as set out in terms of objectives in Annex I of the Directive (which is reproduced in Schedule 2) and as they may be given concrete form in documents (interpretative documents) published in the 'C' series of the *Official Journal of the European Communities*. (The Construction Products Regulations 1991)

ETA guideline
Guideline for European technical approval for a product, or family of products, mandated by the Commission to EOTA, and containing the following:

• list of relevant interpretative documents;
• specific requirements for products within the meaning of the essential requirements;
• test procedures;
• methods of assessing, and judging the results of the tests;
• inspection and conformity procedures;
• period of validity of the European technical approval.
(CPD Art.11: 89/106/EEC)

Eurocode
Structural design code whose drafting, after support by the Commission, has been taken over by CEN (the European Committee for Standardization) as a draft European standard (ENV).

European legislation
Community law consists of:

• that part of international law governing treaties and international institutions;
• EC Treaties and their Annexes;

- secondary legislation by the Community institutions: Regulations; Directives; Decisions; Opinions; Recommendations;
- that part of national law implementing Community provisions.
  (Butterworths Guide to the European Communities)

European standard (EN)

Standard approved by CEN, or by CENELEC (European Committee for Electrotechnical Standardization) as a European Standard, or Harmonization Document (HD), according to the common rules of these organizations: pre-ENs are drafts of European standards issued for comment; ENVs (draft European standards), issued by CEN for trial use for a defined time after which they will become European standards unless problems have arisen during trial use.

European technical approval (ETA)

Favourable technical assessment of the fitness for use of a construction product for an intended use, issued for the purposes of the Directive by a body authorized by a Member State to issue European technical approvals for those purposes, and notified by that Member State to the European Commission. (The Construction Products Regulations 1991)

Favourable technical assessment of the fitness for use of a product for an intended use, based on fulfilment of the essential requirements for the building works for which the product is used. (CPD Art.8(1): 89/106/EEC)

factor of safety

Non-dimensional quantity by which characteristic values are multiplied to achieve a desired minimum reliability over a predetermined range of design variables (H.W. Harrison and E.J. Keeble, *Performance Specifications for Whole Buildings*, 1983)

factory production control

Permanent internal control of production exercised by the manufacturer, whereby:

- all the elements, requirements and provisions adopted by the manufacturer are documented in a systematic manner in the form of written policies and procedures;
- that documentation ensures a common understanding of quality assurance, and enables the achievement of the required product characteristics, and the effective operation of the production control system to be checked; although CPD does not require conformity with the EN 29000: *Quality systems* series, they are referred to in EC SCC Paper 7.

(The Construction Products Regulations 1991; see also CPD Art.13:89/106/EEC; EC SCC Guidance Paper 7: *Guidelines for the performance of factory product control for construction products*)

failure
Termination of the ability of an item to perform a required function.
(BS: 4778:1991 = ISO 8402)

first loss
Alternative to full value insurance, which may or may not be available in particular circumstances. (NEDO, *Build Report*, 1988)

first-party insurance
Policy where payment is first made to the insured, not to a third party.
(NEDO, *Build Report*, 1988)

function
A role for which an item is specially suited. (BS 6100 Part 1.7:1989)

functional requirement
Statement in qualitative terms of a function that a construction or part of a construction must fulfil. (UN ECE Working Party on Building, 1980)

guidance papers
Papers dealing with the implementation and application of the Construction Products Directive (89/106/EEC); but which are not legal interpretations, or are judicially binding.

harmonized European standard (HEN)
Standard mandated to CEN by the European Commission to satisfy certain Essential Requirements described in the New Approach Directives, supplemented if necessary in Interpretative Documents issued by the Commission; produced by a Technical Committee under CEN rules, accepted by the Commission as meeting their mandate, and published in the ECOJ; may be called up as a regulation by a Member State as a condition of placing a product on the market, and under defined circumstances in public procurement; compliance with a harmonized European Norm is mandatory, and hence absolute; enforcement is the subject of Member State regulations.

Liability will be determined by the appropriate court, and will not be affected by contractual obligations between first and second parties. There are three types of harmonized European standards:

- category A standards: fundamental standards related to the design and execution of works, and to the basic data of products, closely linked to relevant essential requirements;
- category B standards: intermediate standards, related to whole families of products and applying to common characteristics of these product families;
- category C standards: standards applying to more or less homogeneous product families or to products which, where applicable, differentiate between defined intended uses.

homologate
To express assent. (*Oxford English Dictionary*)
Act of confirming/ratifying. (Scottish Law)

homologation
Confirmation; a term, unfamiliar to the English construction industry, used in a number of European countries to denote that a technical specification – usually a national standard – has been officially approved and adopted for regulation purposes by a public body, usually a government department.

importer
One who imports into the EC a product for sale, hire, leasing or distribution in the course of business.

inadmissible deformation
Deformation, or cracking of the works or part of the works, which invalidates the assumptions made for the determination of the stability, the mechanical resistance or the serviceability of the works or parts of the works, or causes significant reduction in durability. (EC Interpretative Document for the Essential Requirement: Mechanical Resistance and Stability, 1991)

informative references
Parts of a BS which refer to other publications that provide information or guidance. (BS 0: Part 1 1991)

informative element
That part of a European standard which describes its scope, and any information either a supplier wishes to bring to the attention of the consumer or which the consumer needs to assist his choice of products.

inspection
Process of measuring, examining, testing, gauging or otherwise comparing the item with the requirements. (BS 4778:1991 = ISO 8402)
Primarily concerned with defect detection, and must be planned and carried out by personnel independent of those who perform the work.

inspection body
Impartial body having the organization, staffing, competence and integrity to perform according to specified criteria functions such as assessing, recommending for acceptance and subsequent audit of a manufacturer's quality control operations, and selection and evaluation of products on site, in factories or elsewhere, according to specified criteria.
The standard that should be used as a basis for assessment is EN 45011 pending the adoption by CEN of a specific standard for inspection bodies.

Body designated as an inspection body for the purposes of the Directive by a Member State, and notified by that Member State to the European Commission. (The Construction Products Regulations 1991)

Body having the task of overseeing the manufacturer's quality control and checking its validity. (UEAtc Rule T.03, 1988)

intended use
Role, or roles a product is intended to play in fulfilment of the Essential Requirements. (EC Interpretative Document for the Essential Requirement: Mechanical Resistance and Stability, 1991)

interpretative document
Document prepared by technical committees, in which Member States participate, on the instruction of the European Commission after consultation with the Standing Committee on Construction, which give concrete form to the essential requirements by:

- harmonizing terminology and the technical bases;
- indicating classes, or levels for each requirement where necessary; methods of calculation and of proof, technical rules for project design etc.;
- serving as reference for the establishment of harmonized standards, guidelines for European technical approvals, and recognition of national technical specifications.

Interpretative documents shall be published in the C series of the *Official Journal*. (CPD Ch.IV: 89/106/EEC)

Kitemark
Registered trademark, owned by BSI and only used by manufacturers licensed by BSI under a particular Kitemark scheme; indicates that BSI has independently tested samples of the product against the appropriate British Standard and confirmed that the standard has been complied with in every respect; the manufacturer will also be required to produce and maintain a quality system based on BS 5750: *Quality systems*. (*BSI Buyers Guide*, 1986)

label
Strip of paper, cardboard, metal etc. for attaching to an object, and bearing its name, description or destination. (*Shorter Oxford English Dictionary*)

laboratory accreditation
Formal recognition that a laboratory is competent to perform specific tests, or specific types of test. (ISO/IEC Guide 2)

latent defect
One which exists in the form of a deficiency in the product, but has

not yet caused or turned into damage to the product. (NEDO, *Build Report*, 1988)

latent defects protection insurance
Generic term for all forms of insurance providing indemnity for the costs arising from latent defects, whether in the form of users' latent damage cover, building guarantee or liability cover, and whether renewable or on a long-term basis.

level
Value reached by a quantity (physical property or performance), expressed by means of a number and a unity resulting from a table of generally accepted values, or a calculation or a measurement test. (EC SCC Guidance Paper 11, 1991)

liability insurance
Policy which protects insured against his legal liability to pay compensation or damages to a third party, and the costs of defence. (NEDO, *Build Report*, 1988)

Professional indemnity, public liability or other forms of cover that indemnify a producer who is found legally liable for damage.

load-bearing construction
Organized assembly of connected parts designed to provide mechanical resistance and stability to the works; 'the structure'. (EC Interpretative Document for the Essential Requirement: Mechanical Resistance and Stability, 1991)

loadings
Actions, and other influences which may cause stress, deformations or degradations in the works during their construction and use. (EC Interpretative Document for the Essential Requirement: Mechanical Resistance and Stability, 1991)

local products
Products that correspond to specific local uses or conditions, and which a Member State may adapt to its market for use on its territory without being subject to harmonized technical specifications or the attestation procedures required to obtain the CE mark. A local product cannot carry a CE mark. (Based on EC SCC draft Guidance Note 1)

logo
Popular abbreviation of printing term 'logotype', or a 'type' containing a word, or two or more letters, cast in one piece. (*Shorter Oxford English Dictionary*)

maintainability
The ability of an item (under combined aspects of its reliability, main-

tainability and maintenance support) to form its required function at a stated instant of time or over a stated period of time. (BS 5750)

mandate
A judicial, or legal command from a superior to an inferior. (*Shorter Oxford English Dictionary*)

A command from the European Commission to one of the European Standards Organizations to establish a harmonized European standard, issued after consulting EC SCC. (CPD Art.11: 89/106/EEC)

manufacturer's quality control
Includes:

- quality of various materials and formulations;
- the manufacturing process;
- the quality of the finished product withreference to the Agrément specification.

(UEAtc Rule T.03, 1988)

mark
Sign, indication, brand, characteristic property. (*Shorter Oxford English Dictionary*)

marking
Use of the mark implies a commitment by the manufacturer that the marked product conforms with the technical specification. (UEAtc Rule T.03, 1988)

material damage insurance
Policy which protects insured against damage to his own property, or that for which he is responsible. (NEDO, *Build Report*, 1988)

minor part product
Construction product which is indicated in a list of products which play a minor part with respect to health and safety, drawn up, managed and revised periodically by the European Commission. (The Construction Products Regulations 1991)

mutual recognition
Recognition by UEAtc members of validity of product certification relating to an Agrément which is the subject of an Agrément Confirmation, taking into account the legalprovisions of their country (UEAtc Rule T.03, 1988)

national standards body
Standards body recognized at the national level, that is eligible to be the national member of the corresponding international or regional standards organization. (BS 0: Part 1 1991)

national technical specification
National technical specification which a Member State regards as complying with the Essential Requirements, the text of which is communicated by that Member State to the European Commission, and reference to it published in the *Official Journal of the European Communities*. (The Construction Products Regulations 1991)

negligence
Breach of duty to take reasonable care. (NEDO, *Build Report*, 1988)

no-fault compensation
Compensation available without need to prove fault. (NEDO, *Build Report*, 1988)

normal maintenance
Set of preventive and other measures which are applied to works to enable the works to fulfil all its functions during the working life; measures include cleaning, servicing, repainting, repairing, replacing parts of the works where needed. (EC Interpretative Document for the Essential Requirement: Mechanical Resistance and Stability, 1991)

normative reference
That part of a British Standard which incorporates, by reference, provisions from specific editions of other publications. (BS 0: Part1 1991)

Opinion (EC)
View delivered by Council and/or Commission having no binding force, but merely stating the opinion of the Community institution that issues it on a matter. (DTI: *The Single Market: The Facts*, 1992)

performance
A quantitative expression (value, grade, class or level) of the behaviour of a works, part of the works or product for an action to which it is subject or which it generates under the intended service conditions (for the works or parts of the works) or intended use conditions (for products). (EC Interpretative Document for the Essential Requirement: Mechanical Resistance and Stability, 1991)

performance assessment
Judgement of predicted performance in use based on the comparison of test data with the performance requirement. (BS 6100 Part 1.7:1989)

performance profile
Graphical representation of the performance provided by an item in respect of several performance characteristics (used to compare the overall qualities of alternative solutions). (H.W. Harrison and E.J. Keeble, 1983)

performance requirement
Statement in quantitative terms of an attribute that a construction, or part of a construction, must have in order to fulfil one or more functional requirements. (UN ECE Working Party on Building, 1980)

performance specification
Detailed description which states performance required, and may refer to tests. (UN ECE Working Party on Building, 1980)

performance-aided specification
Specification partly in performance and partly in prescriptive terms. (H.W. Harrison and E.J. Keeble 1983)

performance standard
Acceptance levels for the performance of an item (BS 6100 Part 1.7:1989)

precision (in testing)
The closeness of agreement between selected individual measurements or test results (ASTM, 1963)

producer
Anybody engaged by the client for this building project; in some countries, also the local building authority, any subcontractor, any subconsultant, any public or private laboratory involved in the project (whether or not engaged by the client). (CIB W87)

Manufacturer of a finished product/component, producer of a raw material, a person fixing a trade mark.

All who initiate and bring to fruition building projects, whether private sector, developers, owners building on their own account, or public sector organizations. (NEDO, *Build Report*, 1988)

product
Any industrially manufactured product, and any agricultural product. (Article 1 EC Directive 83/189/EEC as amended by 88/182/EEC)

product approval
Declaration by a body vested with the necessary authority by means of a certificate, or mark of conformity, that a product is in conformity with a state of published criteria.

product standard
Standard which sets down acknowledged technical requirements for a manufactured product that may be incorporated into buildings in either specific or performance terms; if in specific terms, the standard needs to state the purpose of the product and its durability. The way their achievement is measured and assessed may be stated either in the standard, or by reference to a separate standard test or tests. (RIBA Practice Notes, April 1986)

quality

Quality is the sum of:

- knowing customer's needs;
- designing to meet them;
- faultless construction;
- reliable bought-in components and subassemblies;
- certified performance and safety;
- suitable packaging;
- punctual delivery;
- effective back-up services;
- feedback of field experience.

(National Quality Campaign, 1983)

The totality of the attributes of a building which enable it to satisfy needs, including the way in which individual attributes are related, balanced and integrated in the whole building and its surroundings. (BRE Report, *A survey of quality and value in building*, 1978)

quality accreditation

Formal recognition of firms who have adopted systematic procedures in the control of their operations.

quality assessment schedule

A document covering:

- scope of scheme i.e. product type and/or processes covered with any exclusions;
- identification of any defined product specification, test methods, material or component requirements;
- any essential application of the general requirements of the quality system.

quality assurance

All activities and functions concerned with the attainment of quality. (BS 4778:1991 = ISO 8402)

A management process to provide high probability that the objective of product or service will be achieved; it is not a selection process to provide a suitable level of performance in the product or service on offer. (RIBA Practice Notes, April 1996)

A declaration by a supplier that it operates an organizational system capable of consistently achieving the quality required.

quality audit

Independent examination of quality to provide information; can relate to the quality of product, process or system; usually carried out on a periodic basis and involves the independent and systematic examination of actions that influence quality. (BS 4778:1991 = ISO 8402)

quality control
Operational techniques and activities that sustain the product or service quality to specified requirements; also the use of such techniques and activities. (BS 4778:1991 = ISO 8402)

quality management
That aspect of the overall management function that determines and implements the quality policy. (BS 4778:1991 = ISO 8402)

quality management certification
Act of certifying, by means of a certificate of conformity or a mark of conformity, that a supplier's quality management system is in conformity with specific requirements. (NACCB Criteria of Competence, undated)

quality management system
The organization, structure, responsibilities, activities, resources and events that together provide organized procedures, and methods of implementation to ensure the capability of the organization to meet quality requirements. (BS 4778:1991 = ISO 8402)

quality manual
A document setting out the general quality policies, procedures, and practices of an organization. (BS 4778:1991 = ISO 8402)

quality plan
A document derived from the quality programme (extended if necessary) setting out the specific quality practices, resources and activities relevant to a particular contract or project. (BS 4778:1991 = ISO 8402)

quality policy
The overall quality intentions, and direction of an organization as regards quality as expressed by top management. (ISO 8402)

quality procedure
The documents which describe the activities involved in conducting business which are pertinent to the achievement of quality. (CIRIA Report 109)

quality programme
A documented set of activities, resources and events serving to implement the quality system of an organization. (BS 4778:1991 = ISO 8402)

quality system
The organizational structure, responsibilities, procedures, processes and resources for implementing quality management requirements. (ISO 8402)

quality-system certification
Certification, by an independent third party, to internationally accepted standards.

quality in testing
The achievement of reliable, usable results; depends on:

* personnel structure having the necessary fundamental competence;
* knowledge of testing and measurement technology;
* properly maintained and calibrated equipment;
* a simple organization, in which distribution of responsibility and authority is clearly understood;
* a clear and simple administrative system that embraces such aspects as handling of materials, production of reports, filing and confidentiality;
* the right technical and working environment.

recognized national technical specification
National technical specification communicated by a Member State to the Commission, and judged by the Commission, after consulting the EC SCC, as presuming to conform to the Essential Requirements. (CPD Art.4.3: 89/106/EEC)

Recommendation (EC)
Matter made by Council and/or Commission, having no binding force but merely states the view of the Community institution that issues the Recommendation. (DTI, *The Single Market: The Facts*, 1992)

registration of firms of assessed capability
System for the assessment of the capability of a firm to manufacture its products to specifications where there are no suitable British Standards. (*BSI Buyers Guide*)

regulations
Document providing binding legislative rules, that is adopted by an authority. (BS 0: Part 1 1991)

regulation system
Binding system of rules which contain legislative, regulatory or administrative rules, adopted and published by an authority legally vested with the necessary power. (UN ECE Working Party on Building, 1980)

Regulation (EC)
Item of Community legislation directly applicable to all Member States, and which does not have to be confirmed by national parliaments; in the case of a conflict with national law, the Regulation prevails. (DTI, *The Single Market: The Facts*, 1992)

regulatory body
Governmental, or non-governmental body which has legal power to enforce a technical regulation. (UN ECE Working Party on Building, 1980)

relaxation
Specific relaxation, for a particular building project, from one or more statutory requirements. (UN ECE Working Party on Building, 1980)

relevant national standard
A national standard of which a reference is published:

- in the United Kingdom, by the Secretary of State in such a manner as he considers appropriate; or
- in another member State;
- and which corresponds to a harmonized standard the reference to which is published in the *Official Journal of the European Communities*. (The Construction Products Regulations 1991)

relevant technical specification
A European technical approval. a national technical specification, or a relevant national standard. (The Construction Products Regulations 1991)

Safety Mark
BSI mark appearing on a number of products which conform to British Standards specifically concerned with safety, or to the safety requirements of standards which cover other product characteristics as well. (*BSI Buyers Guide*)

self-certification
Conformity certification in which one, or more suppliers are responsible for conformity certification of their products or services; there may be initial third-party type testing of the product or assessment of the service, but responsibility for declaration of conformity and maintenance of quality rests with the supplier.

serviceability limit state
State beyond which specified criteria for the structure related to its use, or function are no longer met. (EC Interpretative Document for the Essential Requirement: Mechanical Resistance and Stability, 1993)

specification
Document which prescribes in detail the requirements to which supplies, or services must conform. (BS 4778:1991 = ISO 8402)

Communicates the requirements, or intentions of one party to another, providing the basis of agreement; a detailed statement of the objectives to be satisfied by the product, or service concerned and, where appropriate, of the procedures for verification. (BS 5750 Part 4)

specific requirement
Statement of a particular specified way in which a construction, or part of a construction should be designed or executed. (UN ECE Working Party on Building, 1980)

standard
A technical specification approved by a recognized standardization body for repeated, or continuous application, with which compliance is not compulsory. (Article 1 EC Directive 83/189/EEC as amended by 88/182/EEC)
  Document, established by consensus and approved by a recognized body, that provides for common and repeated use, rules, guidelines, or characteristics for activities, or their results, aimed at the achievement of the optimum degree of order in a given context; standards should be based on the consolidated results of science, technology and experience, and aimed at promotion of optimum community benefit. (BS 0: Part 1 1991, based on BS EN 45020: 1991)

standards body
Standardizing body recognized at national, regional or international level whose principal function, by virtue of its statutes, is the preparation and/or publication of standards, and/or approval of standards prepared by other bodies. (UN ECE Working Party on Building, 1980)

standardization
Activity of establishing, with regard to actual or potential problems, provisions for common and repeated use, aimed at the achievement of the optimum degree of order in a given context. (BS 0: Part 1 1991)

state of the art
Developed stage of technical capability at a given time as regards products, processes or services, based on the relevant consolidated findings of science, technology and experience.

Statutory Instrument
Measure having legal force, used by the responsible government department, usually on the authority of a Statute. (Legal)

Statute Law
Laws of general application enacted by or on the authority of Parliament. (Legal)

supply
Offering to supply; agreeing to supply; exposing for supply; and possessing for supply, and cognate expressions shall be construed accordingly. (The Construction Products Regulations 1991)

surveillance programme
Programme of visits by an accreditation or certification body designed to establish continuing requirements of the accreditation or certification scheme.

technical control bureau
Body providing technical design checks; *bureau de contrôle* [F]. (UN ECE Working Party on Building, 1980)

technical regulation
Regulation, or part of a regulation, which consists of or refers to a technical specification, including the applicable administrative provisions. (UN ECE Working Party on Building, 1980)

technical specifications
Totality of the technical prescriptions contained in particular in the tender documents, defining the characteristics required of a work, material, product or supply, which permits a work, material, product or supply to be described in a manner such that it fulfils the use for which it is intended by the contracting authority; these technical prescriptions shall include levels of quality, performance, safety or dimensions, including the requirements applicable to the material, the product, or to the supply as regards quality assurance, terminology, symbols, testing and test methods, packaging, marking or labelling. They shall also include rules relating to design and costing, the test, inspection and acceptance for works and methods or techniques of construction and all other technical conditions which the contracting authority is in a position to prescribe, under general or specific regulations, in relation to the finished works and to the materials or parts which they involve. (Article 1 EC Directive 83/189/EEC as amended by 88/182/EEC)

test
Inspection process in which a functional requirement is measured or observed and usually in which stress or energy is applied to the item. (BS 4778:1991 = ISO 8402)

testing agency
Testing laboratory acting for a certification body to carry out the testing, and other associated activities necessary for conformity certification as requested by the certification authority. (UN ECE Working Party on Building, 1980)

testing laboratory
Laboratory which measures, examines, tests, calibrates or otherwise determines the characteristics, or performance of materials or products. (EEC draft)

The standard that should be used as a basis for assessment is EN 45001: *General criteria for the operation of testing laboratories.*

Impartial body, which may be a public agency or a private organization or individual, possessing the necessary competence and reliability to operate testing services. In the UK, NAMAS, the National Measurement Service of the National Physical Laboratory, operates on behalf of the UK Government the National Laboratory Accreditation Scheme, under which accreditation certificates are awarded to laboratories undertaking testing work detailed in a NAMAS schedule.

third-party certification
Certification carried out directly by an accredited certification body, or under its surveillance. (*A National Strategy for Quality*, 1978)

total quality management
A way of managing an organization which aims at the continuous participation and cooperation of all of its members in the improvement of:

- the quality of its products and services;
- the quality of its activities;
- the quality of its goals;
- to achieve customer satisfaction, long-term profitability of the organization and the benefits of its members, in accordance with the requirements of society. (Draft addendum to ISO 8402: *Quality vocabulary*)

traceability (of test results)
The property of a result of a measurement whereby it can be related to appropriate measurement standards, generally international or national standards, through an unbroken chain of comparisons. (*Vocabulaire International de Métrologie*)

transposition (of a Directive)
Transference, into national legislation of a Member State, result to be achieved in a binding EU Directive, but leaving form and methods of its achievement to the Member State.

type approval
Approval of a certain product, or group of products, considered by the approval body as representative for the continuous production. (UN ECE Working Party on Building, 1980)

UEAtc guide
Common technical platform on the basis of which a technical Agrément, issued by an institute of UEAtc, can be readily confirmed by another institute; i.e. recognized as equivalent by another institute,

as all UEAtc institutes have agreed to the context of these guides. (UEAtc Information 25, January 1992)

ultimate limit state
State close to structural failure. (EC Interpretative Document for the Essential Requirement: Mechanical Resistance and Stability, 1991)

users' latent damage insurance
Policy in the name of the building owner, or tenant, and assignable to his successors, which directly indemnifies the building user for the cost of insured types of damage.

works
Construction works, including both buildings and civil engineering works. (The Construction Products Regulations, 1991)

16.2 FRENCH

In this section terms used in French regulations, standards and codes, quality assurance, testing and certification are listed. The gender of nouns is noted to ensure correct use in the French language: [f] feminine; [m] masculine.

abrogation [f]
Repeal.

appel des offres [m]
Call for tenders.

appréciation technique d'expérimentation (Atex) [f]
A preliminary assessment of innovatory products and techniques by CSTB, restricted to use on demonstration or experimental schemes.

arrêté [m]
Ministerial order; local regulation, in the form: le ministre délégué.

assurance CRC [f]
Insurance covering for three years risks of not satisfying regulations.

assurance de la qualité [f]
Quality assurance.

assurance dommage [f]
Damage insurance.

assurance responsabilité décennale [f]
Insurance covering ten-year responsibilities of participants in construction work.

avis technique [m]
A certificate issued by CSTB giving an assessment of likely performance and fitness for use, and an account of the reasons on which the assessment is based. The literal translation is 'technical opinion' or 'technical advice'.

avis technique agré [m]
Acceptance of an *avis technique* by a technical committee of French insurers to indicate that a product or technique has normal cover in damage and responsibility insurance. It then takes on more or less the older *agrément* status.

bureau de contrôle technique agrée [m]
Independent technical control office accredited by government.

bureau d'études techniques (BET) [m]
Technical design office; independent or associated with a contractor's organization.

cahier de charges [m]
Conditions of contract.

cahier des clauses administratives générales (CCAG) [m]
General specification clauses in public contracts.

cahier des clauses administratives particulières (CCAP) [m]
General specification clauses in private contracts.

courier d'entreprise à distribution exceptionelle (CEDEX) [m]
Special arrangement for collecting an organization's mail.

certificat de conformité [m]
Certificate of conformity to rule/standard.

certificat de qualification [m]
Certificate given by an accredited certification body certifying certain characteristics of a product or technique.

certification par un tiers [f]
Third-party certification.

certification de la gestion de la qualité [f]
Quality management certification.

circulaire [f]
Ministerial circular published in the *Official Journal* in the form *le ministre délégué*.

classement d'immeubles [m]
'Listing' of buildings.

classement des sites [m]:
'Listing' of historic sites.

code civil (CC) [m]
Principal collection of laws and regulations relating to civil matters.

code de la construction et de l'habitation (CCH) [m]
Collection of laws and regulations relating to construction and housing.

code des devoirs professionnels [m]
Code of professional conduct.

code des marchés publics [m]
Code (rules) for public contracts.

code d'urbanisme [m]
Planning code.

collectivités locales [f]
Local authorities (*communes*).

conception assistée par ordinateur (CAO) [f]
Computer-aided design.

conducteur de travaux [m]
Foreman-in-charge.

conjoncture [f]
Prediction based on relevant facts.

conseil juridique [m]
Solicitor.

Conseil d'État [m]
State Council (supreme administrative court; consulted and rules on subsidiary legislation).

contrôle par sondage [m]
Spot-checks.

contrôleur technique agré [m]
Technical control office or person licensed by a government commission to examine designs, inspect sitework, appraise risks and issue certificates of acceptance work on behalf of a building owner within arrangements for civil code liabilities.

Cour de Cassation [f]
Supreme appeal court for civil cases.

déclaration de conformité [f]
Self-certification.

décret [m]
Decree; ordinance: in France published in the *Official Journal* under the signature of *Le Premier ministre*.

Direction départmentale de l'équipement (DDE) [f]
Agency of construction ministry in a Département.

dérogation [f]
Dispensation; relaxation.

devis [m]
Estimate; specification; bill of quantities.

document technique unifié (DTU) [m]
Technical code of practice, prepared by an industry committee under the direction of CSTB.

dommage-ouvrage-assurance [f]
(Building) damage indemnity insurance.

exigences essentielles [f]
Essential requirements.

établissement public [m]
Public executive agency set up by government decree; for example, for the new national library project.

établissement public à caractère commercial et industriel [m]
Public executive agency set up by government decree to carry out research and similar activities, including work for payment; for example CSTB, the building research centre.

établissement reçevant du public (ERP) [m]
Building open to the public; in fire safety legislation there are a number of ERP classes.

exigence fonctionelle [f]
Functional requirement.

expert [m]
Technical assessor; expert witness.

fournisseur [m]
Supplier.

géomètre [m]
Land surveyor.

génie civile [m]
Civil engineering.

gestion de la qualité [f]
Quality control.

gros-oeuvre [f]
Main structural work in a building, the 'carcass'.

immeuble [m]
Block of flats, or offices.

infraction [f]
Breach of law, of duty.

ingénierie [f]
Engineering design work.

label de qualité [m]
Quality label.

label 'qualitel' [m]
Label given by Association Qualitel to new dwellings, particularly flats which confirm certain features in their design evaluated independently.

laboratoire agré [m]
Approved testing laboratory.

maire [m]
Mayor; as first official of a local authority (*commune*) a mayor, or his deputies in a larger authority (*ville*), have executive and not just ceremonial responsibilities.

mairie [f]:
Administrative offices of a local authority.

maître d'oeuvre [m]
Architect or surveyor responsible for design and control of building works.

maîtrise d'oeuvre [f]
Design and control process, project organization.

maître d'ouvrage [m]
Client; building owner.

maîtrise d'ouvrage [f]
Ownership of a building, client organization.

mandat [m]
Commission.

marché de l'état [m]
Government contract.

marché d'étude [m]
Design contract.

marché publique [m]
Public contract.

marché à prix ferme [m]
Fixed-price contract.

marque de conformité [f]
Mark of conformity.

marquage CE [f]
CE mark.

marque 'A' [m]
Mark available to French accredited certification bodies.

marque NF [m]
Conformity mark given by AFNOR, the French standards organization, to products; similar marking arrangements exist in the UK, e.g. the BSI Kitemark.

menus-oeuvrages [m]
Those services and finishes of a building subject to three-year damage responsibilities.

métreur [m]
Quantity surveyor employed by the contractor.

métreur-vérificateur [m]
Independent quantity surveyor.

norme [f]
Standard, usually limited to physical properties and their testing.

norme européene harmonisée [f]
Harmonized European standard (HEN).

norme française (NF) [f]
Standard issued by AFNOR, the French national standards organization.

organisme [m]
Body, legal or administrative entity.

organisme à activités normatives [m]
Standardizing body.

organisme certificateur agré [m]
An officially approved (accredited) certification body.

organisme de certification [m]
Certification body.

organisme de contrôle [m]
Inspection agency.

organisme d'inspection agré [m]
An officially approved (accredited) inspection body.

organisme de normalization [m]
Standards body.

ouvrage de clos et couvert [m]
'Weathershield': i.e. that part of building the subject of decennial responsibilities by constructors.

permis de construire [m]
Building and planning permit issued from *mairie*.

personne morale [f]
Corporate body.

police dommages-ouvrage [f]
Insurance policy taken out by building owner, in some cases obligatory, to ensure rapid repair of any damage covered by Civil Code decennial responsibilities.

police responsabilité décennale [f]
Insurance policy taken out by individual participants in construction
– architect, structural engineer or design office, contractor etc. – to
cover ten-year Civil code responsibilities.

police unique de chantier [f]
Decennial damage insurance policy offering a building owner complete
protection by grouping under one site policy the *dommages-assurance*
and *responsabilité décennale guarantees*.

pouvoirs publics [m]
The authorities; the Administration.

préfet, commissaire de la République [m]
Senior central government official at level of *département* or *région*.

prescription de performance [f]
Performance requirement.

prestation [f]
Supply.

prestations intellectuelles [f]
Supply of intellectual services.

prestations de fournitures [f]
Supply of equipment.

prestations de services [f]
Supply of services.

projecteur de bâtiment [m]
Building designer.

qualité [f]
Quality; degree of excellence; also property.

réception des travaux [f]
Taking delivery, acceptance of completed works.

**Recueil des eléments utile à l'établissement et à l'exécution des
projects et marchés de bâtiment en France (REEF) [m]**
Eleven-volume collection of French legislation, standards, DTUs, technical calculations, examples of product certificates, and avis techniques,
published by CSTB; also available on CD and in a shorter three-
volume edition.

recours en cassation [m]
Appeal to courts.

référence aux normes [f]
Reference to standards.

référence aux normes avec identification rigide [f]
Reference to standards by exact identification.

référence aux normes avec identification glissante [f]
Reference to standards by undated identification.

régie [f]
Administration, management.

règle d'exécution [f]
Workmanship code of practice.

règle de calcul [f]
Design (calculation) code of practice.

règlement [m]
Regulation.

règlement nationale [m]
National regulation equivalent to a Statutory Instrument.

réglementation de la construction [f]
Building regulations.

renvoi aux normes [m]
General reference to standards.

réputé satisfaire au règlement
Deemed-to-satisfy provision.

responsable agrée des travaux [m]
Responsible site supervisor.

solidaire [adj.]
Joint (responsibility).

sous-traitant [m]
Subcontractor.

spécification de rendement [f]
Performance specification.

spécification normale [f]
Prescriptive specification.

système de certification [m]
Certification system.

système de certification par une tierce partie [m]
Third-party certification system.

tribunal de grand instance [m]
County court.

utilisateur [m]
User.

vérificateur [m]
Quantity surveyor employed by architect to check sitework.

16.3 GERMAN, DANISH, DUTCH AND SWEDISH

In this section terms in common use in the German language for standards, quality assurance, testing and certification are listed, together with a limited list of similar terms in Danish (Dk), Dutch (NL) and Swedish (S).

To help with the correct use of the German language, the gender of words is given: [f] feminine; [m] masculine; [n] neuter.

aanbeveling [NL]
Recommendation.

aannemer [NL]
Contractor.

aansprakelijkheid [NL]
Responsibility, liability.

afstemming [NL]
Rejection.

Allgemeine Anforderungen an die Bauausfürung [f]
General requirements for building execution.

Allgemeine Grundsätzen [m]
General principles (of a Standard).

Allgemeine Vorschriften [f]
General provisions in a *Land* building ordinance.

ambtelijk [NL]
Official.

Amtliche Prüfstelle [f]
Official testing laboratory.

Anerkannte Prüfanstalt für das Zulassungwesen [f]
Accredited testing laboratory for tests for the *Zulassung* approval procedure.

Anerkennung [f]
Recognition.

Angestellte [m]
Technical employee in a building authority.

Anhang [m]
Appendix; annexe.

Anlage (bauliche Anlage) [f]
Building installation.

ansvarlige myndighed [Dk]
Responsible authority.

ansøgning [Dk]
Application.

Anwendungsverbindlichkeit [f]
Binding applications.

Arbeitgeber [m]
Employer.

Arbeitnehmer [m]
Employee.

attest [NL]
Certificate of fitness for specified use, issued by a body accredited by RvC, the Dutch Council for Certification.

attest-met-certificaat [NL]
Indication of conformity with the relevant Dutch standard as well as an assessment of fitness for use.

attestering af overensstemmelse [NL]
Certificate of conformity.

Aufsichtssbehörde [f]
Regulatory body.

Auftraggeber [m]
Client.

Auftragnehmer [m]
Contracting party, supplier.

Ausbaugewerbe [n]
Auxiliary trades.

Ausführungsanweisung [f]
Carrying out instructions.

Ausnahme [f]
General or partial exemption in statutory requirements

Ausschuss [m]
Committee, board.

Bauassesor [m]
Engineer in high technical service who has passed the 'state examination'.

Bauaufsichtliches Genehmigungsverfahren [n]
Building control procedures to check design conforms to stated requirements, and construction to approved design.

Bauaufsichtliche Richtlinien [f]
Guidelines issued by the highest building authority (*Oberste Baubehörde*) in a *Land* to be observed by a lower authority (*Untere Baubehörde*).

Bauaufsichtsbehörden [f]
Building supervisory (control) authority: see *oberste Baubehörden*.

Baubehörde [f]
Building authority of a State (*Land*), usually under the State building ministry, the title of which varies from *Land* to *Land*; authority may function at a number of levels, with powers delegated to a larger municipality (*Stadt*).

Baugenehmigung [f]
Building permit.

Baugesetz [n]
In Federal Switzerland, a Canton enacts a *baugesetz*, which a municipality may or may not supplement by a more detailed *bauordnungen*.

Baugesetzbuch [n]
Collected volume of building laws.

Bauherr [m]
Building owner; client.

Bauleistungen [f]
Building work.

Baunutzungsverordnung (BauNVO) [f]
Land use ordinance.

Bauordnungen [f]
Building ordinance (law) of a *Land*, covering administration, issue of building permits, inspection, enforcement, waivers and suitability of products and systems. In Federal Germany, the *Länder* are the higher legal authorities for building control, administered through a building ordinance on a model (*musterbauordnung*). There are provisions in the ordinance for general or partial exemption from statutory requirements (*ausnahme*); and granting of building permits (*ausgenehmigung*).

Bauprodukten-Richtlinie – Richtlinie des Rates von 21 Dezember 1988
89/106/EEC the Construction Products Directive.

Bautechnische Prüfungsverordnung (BauPrüfVO) [f]
Land ordinance relating to the work of checking by independent licensed structural engineer (Prüfingenieur).

Bauträger [m]
Builder.

Bauüberwachung [f]
Inspection of building work, usually on site.

Bau- und Prüfgrundsätzen [m]
Principles for building and testing.

Bauverbot [n]
Building restriction.

Bauvorhaben [n]
Building project.

Bauvorlagverordnung (BauVorlVor) [f]
Form of building application regulations.

Bauwesen [n]
Building, building industry.

Befreiung [f]
Specification dispensation from legal requirements, usually for one building.

beleid [NL]
Policy.

bepalingen [NL]
Conditions; provisions.

bericht [NL]
Report.

beroepsvereniging [NL]
Trade association.

besluit [NL]
Decree.

bestek [NL]
Specification.

bestemmelser [Dk]
Administrative provisions.

bestuur [NL]
Board of management.

bewaking [NL]
Surveillance.

boligministeriet [Dk]
Housing ministry.

bouwbedrijf [NL]
Building industry.

bouwbesluit [NL]
Building decree.

bouwtoezicht [NL]
Building supervision, inspection.

bouverordening [NL]
Municipal building regulations, byelaws; usually based on a model issued by VNG (Association of Dutch Municipalities).

Brandschutz [m]
Fire protection.

byggevarer [Dk]
Construction product.

bygge- og boligstyrelsen [Dk]
Building and housing agency (of Danish housing ministry).

byggnorm [S]
Building regulations, not national standards (like BSs and DIN), though in many cases have a similar influence on building quality.

bygningsreglement [Dk]
Building regulation.

certificaat [NL]
Conformity certificate to Dutch Standards, issued by a body accredited by RvC (Dutch Certification Council).

certificatie-instelling [NL]
Certification authority.

certificerinsgslicens [Dk]
Licence (to use DS mark).

certificerinsgsmaerket [Dk]
(DS) product certification mark

Deutsches Institut für Bautechnik (DIBt) [n]
German Institute for Building technology, an *Ahnalt des öffentlichen Recht*, jointly authorized by the Federal and Länder governments as the coordinating body for the approval of construction products under *Landesbauordungen*.

Durchführungsverordnung [f]
Implementation ordinance.

EF-maerket [Dk]
CE mark.

Eigenüberwachung [f]
Factory quality control by manufacturer.

Eigenschaft [f]
Characteristic.

einheitliche technische Baubestimmungen (ETB) [f]
'Uniform technical building rules', standards for essential building requirements under *Länder* building ordinances; lists of ETBs are appended to ordinances and used by building control authorities as basis for assessment and inspection of construction work.

einheitliches Uberwachungszeichen [n]
Uniform quality certification mark, certifying the legally required quality control in the construction field.

einwandfreie Herstellung [f]
Perfect manufacture.

Empfehlung [f]
Recommendation.

entsprechenenden technischen Regeln [f]
Corresponding (related) technical rules.

erklaering [Dk]
Declaration.

Erschütterungsschutz [m]
Protection against vibrations.

Europaeisk technisk godkendelse [Dk]
European technical approval.

Europese technische goedkeuringen [NL]
European technical approval.

fabrikationstedier [Dk]
Place of manufacture.

Fachkommission Baunormung [f]
Expert committee for building standards.

Fachnormenausschuss Bauwesen [m]
Building council of DNA.

faglige udvalg [Dk]
Working group.

Feuerungsverordnung (FeuVO) [f]
Fire precautions regulations.

forbruger [Dk]
User; consumer.

Fremdüberwacher [m]
Officially authorized inspection body.

Fremdüberwachung [f]
Third-party quality control supervision, e.g. by an independent laboratory or control association (*fremdüberwacher*).

Gefahrenschutz [m]
Safety (protection against hazards).

gemeente [NL]
Municipality.

Genehmigung [m]
Licence; approval; permit.

genehmigungsbedürftige
Subject to authorization.

Geschäftshausverordnung (GhVO) [f]
Regulations for business premises.

Gewerbeordnung [f]
Federal 'health and safety at work' ordinance.

godkendelse [Dk]
Product approval.

Gütenachweis [m]
Any required procedure to prove conformity with requirements, e.g. with those set out in a standard; quality assurance.

Gütesicherung von Baustoffen und Bauteilen [f]
Quality control of building materials and elements, consisting of internal and external (third-party) control, undertaken by quality control associations of manufacturers under supervision of RAL (*Ausschuss für Lieferbedingungen und Gütesicherung*), the Committee for technical conditions (terms of delivery) and quality control.

Güteschutzgemeinschaft [f]
Literally 'quality protection' association: each industry association has its own quality mark (*Gütezeichen*), which is registered with the German Patent Office and is a legally protected trade mark.

Güteüberwachung [f]
General term for quality control procedures.

Gütezeichen [n]
Authorized quality mark.

Hochhaus-Rechtlinien [f]
High-rise buildings 'guidelines'.

Inspektor [m]
Technical engineer who has qualified for the middle technical administrative service.

instellingen [NL]
Institution.

instemming [NL]
Agreement.

inzage [NL]
Inspection.

keuringsintituten [NL]
Testing establisment.

kommunale bestyrelsen [Dk]
Muncipal council.

kommunale myndigheder [Dk]
Muncipal authorities.

Kreisbehörde [f]
District authority.

kwaliteitsverklaringen [NL]
Quality declaration issued by certification body; either *attest*, *certificaat*, or *attest-met-certificaat*.

Land [n]
State; since unification there are 16 *Länder* in the Federal Republic of Germany (*Bundesrepublik*).

Ländersachverständigen-ausschuss (LSA) [m]
Länder expert committee.

Landesbauordnung (LBO) [f]
State building ordinance.

lov [Dk]
Law.

medlemsstaterne [Dk]
Member state.

Musterbauordnung [f]
Model building ordinance prepared by a national committee; has in itself no legal status but serves as basis of State (*Land*) building ordinances.

Muster-Prüfzeichenverordnung ausschuss [m]
Model approval marks regulations.

myndighed [Dk]
Authority.

naamloze vennootschap (NV) [NL]
Limited liability company.

Nachweis der konformität [m]
Verification (proof) of conformity.

nevenbedrijven [NL]
Coodination committee.

Öberste Baubehörde [f]
The 'highest building authority' of a *Land* under the responsible 'construction' ministry; may function at a regional (*Regierungsbezirk*) level.

öffentlich-rechtliche
Subject to public law.

ondernemers [NL]
Owners.

oplysninger [Dk]
Information.

oprichting [NL]
Establishment.

overeenkomst [NL]
Agreement.

overgansperiode [Dk]
Transitional period.

overzicht [NL]
Summary.

privat-rechtliche
Subject to private law.

produktkeuring [NL]
Product inspection, assessment.

prøvningsbetingelser [Dk]
Testing conditions.

Prüfamt [n]
Official public works testing station of a *Land*.

Prüfamt [n] für Baustatik
Official office for checking structural designs.

Prüfingenieur [m] für Baustatik
Consultant structural engineer licensed by a *Land* building ministry to check structural designs in building permit applications.

Prüfstelle [f]
Building materials testing laboratory.

Prüfstelle E (Eigenüberwachung) [f]
A manufacturer's laboratory, or a laboratory under contract to a manufacturer, carrying out factory production control.

Prüfstelle F (fremdüberwachende Stelle) [f]
A recognized laboratory which has been accredited for performing quality control of building products and for issuing the uniform quality certification mark meeting minimum requirements under *Land* building ordinances.

Prüfzeichen [n]
Testing mark; approval mark.

Prüfzeichenverordnung [f]
Approval marks regulations.

prüfzeichenpflichtige Baustoffen [m]
Building products requiring a quality control (approval) mark.

Qualität [f]
Quality; grade; type.

Qualitätsanforderungen [f]
Approval activities and quality requirements.

Raumordnungsgesetz [n]
Federal legislation relating to regional planning.

Referendar [m]
Diploma engineer serving a period of training in the higher technical service before taking the 'state examination'.

regelgeving [NL]
Rule giving; code-making.

Richtlinien [f]
'Guidelines', approximating to codes of practice; may be issued by a professional association like VDI. UEAtc Directives are translated as *Richtlinien*.

Rohbau [m]
Main structural work or the 'carcass' of a building.

saerlige bestemmelser [Dk]
Special requirements.

Stående Byggeudvalg [Dk]
EU Standing Committee for Construction.

standaardbestek [NL]
Standard specification.

stedebouwkundigen [NL]
Town planning.

Stichting [NL]
Foundation; non-profit-making association.

technischer Verwaltungsdienst [m]
Higher technical administrative service.

Tiefbau [n]
Work below ground, i.e. civil engineering.

tilnaermelse [Dk]
Approximation.

tilverkningskontroll [S]
Production control – assessment of a manufacturer's quality system followed by surveillance of his works – operated in association with arrangements for type approvals under the Swedish building regulations; may be operated through the Swedish national testing establishment or an industry (trade) control association.

toelating [NL]
Admission.

toetsen [NL]
Test.

Typengenehmigung [f]
Type approval given for standard (prefabricated) building structures all constructed on the basis of the same design and execution

typgodkännanden [S]
Type (general) approvals for products, systems etc. given by Boverket, the Swedish National Board of Housing, Physical Planning and Building under the Swedish building regulations (Svensk Byggnorm).

Übergangsvorschriften [f]
Provisional regulations.

Überwachung [f]
Inspection.

Überwachungsverein [m]
Approved manufacturers' quality control association.

Überwachungsverordnung (ÜVO) [f]
Quality control regulations.

untere Baubehörde [f]
'Lower' building authority competent for the implementation of the building control procedures, usually at the level of a *Landkreis* (district council) or that of a larger town.

vaesentlige krav [Dk]
Essential requirements.

verantwortliche Bauleiter [m]
'Responsible building supervisor'.

Veringungsordnung für Bauleistungen (VOB) [f]
Set of standards relating to public contract administration, also used in private contracts (DIN 1960-1); published by DIN on behalf of Deutscher Veringungsausschuss für Bauleistungen.

Verordnung [f]
Regulation, order.

Verordnungsmächtigung [f]
Regulating authority.

Versammlungsstättenverordnung [f]
Assembly buildings regulations.

Verwaltungsgeböhrenordnung [f]
Administrative fees regulations.

Verwaltungsrat [m]
Supervisory council.

Verwaltungsverfarhren [n]
Administrative procedures.

Verwaltungsvorschriften [f]
Administrative rules, regulations.

Verzicht [m]
Waiver, disclaimer.

Vorbescheid [m]
Preliminary decision.

voorlicting [NL]
Guidance.

Vorschriften [f]
Direction; order; regulation.

Vorschriftensammlung [f]
Published collection of ordinances and associated legal documents.

wijziging [NL]
Alteration; amendment.

woordvoerder [NL]
Spokesman.

zaak [NL]
Affair, matter.

Zulassung [f]
Permit, 'admission'; a general licence, since 1968, issued by the German Institute for Building Technology (Deutsches Institut för Bautechnik) on behalf on State (*Land*) building ministries, assessing the suitability of products against *Länder* building regulations.

Reference sources 17

There are many sources of information on construction quality and quality standards – international, European and national. A number of databases carry standards information on line, or on CD-ROM discs. European and national organizations listed provide information relevant to their particular sectors in construction. Some services are free; others make a charge. Some restrict use to subscribers. Some professional institutions, research and standards bodies, and specialist publishers and bookshops issue catalogues or booklists – usually annual, and available free. The standards databases to which the BSI Library has access are listed first. Publication series issued by European, UK and other national organizations, and a selected list of annuals and reference books, follow.

17.1 STANDARDS DATABASES

The BSI Library has access to the following databases:

* ISONET, the network of national standards information centres, set up under ISO;
* STANDARDLINE, the BSI on-line database covering all current British Standards, amendments and drafts for comment [since 1985 on the host PFDS (Pergamon Financial Data Systems) and on Fiz Technik];
* DIT, the German standards database, which includes information on legislation and regulations (on hosts PFDS and Fiz Technik);
* NORIANE, the French standards database, which also includes information on regulations (on host Fiz Technik);
* IHS International Standards and Specifications, with references to US military, federal and industry standards; also standards issued by ISO and IEC, CEN and CENELEC, and many national standards, particularly of countries on the Pacific rim (on host Dialog);
* CD-ROM PERINORM. There are two PERINORM multinational, multilingual bibliographic database services on CD-ROM: PERINORM Europe, covering ISO, IEC, CEN and CENELEC international standards and national standards for Austria, France, Germany, the Netherlands, Switzerland and the UK, plus French and German technical regulations; and PERINORM International,

covering additionally ASTM, IEEE, UL and ANSI North American standards.

BSI Library document LIB 002, *A brief guide to standards databases*, gives details of on-line and CD-ROM databases covering standards and associated information. For details, telephone 0181 996 7000.

The BSI Library reports on standards information on-line services in the BSI Standards periodical *INSTEP* (International News on Standards and Exporting). For information on these and other electronic matters, there is an Electronic Products Help Desk at BSI Chiswick: telephone 0181 996 7333; fax 0181 996 7047.

17.2 INTERNATIONAL AND GENERAL EUROPEAN INFORMATION SOURCES

The quarterly catalogue of the Office for Official Publications of the European Communities, L-2985, Luxembourg, lists bookshops from which these publications can be obtained.

17.3 NATIONAL INFORMATION SOURCES

17.3.1 Federal Germany

IRB
Informationzentrum RAUM und BAU des Fraünhofer Gesellschaft

Maintains two databases, in German: BAUFOU, which contains references to ongoing and completed research projects in the fields of building, housing and related aspects of urban planning in German-speaking countries; and FORS, on regional and town planning and associated subjects for Bundesforschungsanstalt für Landeskunde und Raumordnung, Bonn; also manages ICONDA.

DIBt
Deutsches Institut für Bautechnik

Reichpietschufer 74–76, D-1000 Berlin 30

Publishes *Mitteilungen Institut für Bautechnik*, bimonthly bulletin in German giving technical reports mainly on construction materials, German standards and generally recognized rules as technical building regulations issued by higher building authorities, also *Zulassungen* issued by DIBt.

Curt R. Vincentz Verlag

Postfach 62 67, D-3000 Hannover

Publishes *Bauverwaltung*, the official monthly journal of the higher technical construction service, in which are listed new and draft NABau Normen (German building standards), copies of which can be purchased from Beuth-Verlag, Postfach 11 45, D-1000 Berlin 30, as well as news on technical regulations.

17.3.2 France

CSTB Service Publications

84 avenue Jean Jaurès, BP 2 Champs sur Marne, F-77421 Marne la Vallée Cedex 2
Tel: +33 16 (1) 40 50 28 28; fax: +33 16 (1) 45 25 61 51

Issues an annual catalogue, *Publications du CSTB et du Plan Construction et Architecture*, listing: REEF – Collection of *Documents Techniques Unifiés* (codes and standard specifications); *Normes applicable au Bâtiment: Législation du Bâtiment*; *Règles de Calcul et Examples des Solutions*; *Science du Bâtiment*; *Cahiers du CSTB*; *CSTB magazine*; and, for an additional subscription, *Bulletin mensuel des Avis Techniques*; and many other technical and scientific reports. There is also information on CD-REEF, a new development that carries the 16 5000 pages of REEF, including tables and figures, and is replaced by an update disc by subscription four times a year.

Publications du Moniteur

17 rue d'Uzès, F-75002 Paris

Publishes weekly magazine *Le Moniteur*, with supplement of construction legislation, standards etc.; also Editions du Moniteur catalogue, annual list of publications, mainly in French, on architecture and building, also construction legislation.

Qualité Construction

36 rue de Picpus, F-75012 Paris

Publishes *Syncodés Information*, a quarterly publication in French devoted to construction, insurance and maintenance.

17.3.3 Netherlands

Stichting Bouwkwaliteit

Treubstraat 1, NL-2288 EG Rijswijk (ZH)

Publishes *Bouwkwaliteit-Nieuws*, periodical newsletter in Dutch with brief English summaries, giving details of European construction developments in the Netherlands, with lists of product assessment schedules issued by Harmonisatie Commission Bouw.

17.3.4 Sweden

BYGGDOK
Institutet för byggdokumentation (Swedish Institute of Building Documentation)

Hälsingegatan 47, S-113 31 Stockholm

Swedish Council for Building Research

Sankt Göransgatan 68, S-112 33 Stockholm

Publishes annual list of published research work, generally in English; and synopses of Swedish building research, English version of periodical list of recent research reports with short synopses. Priced reports listed, most in Swedish but some in English, are usually available from AB Svensk Byggtjänst, S-171 88 Solna.

17.3.5 Switzerland

CRB
Swiss Centre for Rationalization in Building and Civil Engineering

Zentralstrasse 163, CH-8003, Zürich

Annual list of publications, mainly standard building descriptions and cost analyses in French and German.

Schweizer Baudokumentation

CH-4249 Blauen

Publishes *Docu Bulletin*, periodical issued ten times a year in French and German, giving information on architecture and construction; also lists of books and new products.

17.3.6 United Kingdom

BRE Bookshop, Building Research Establishment, Garston, Watford WD2 7JR
Tel: 01923 664444; fax: 01923 664400

BRE Bookshop Catalogue: annual classified list of Building Research Establishment publications and other information products; updated in *BRE News of Construction Research* five times a year.

BSI Standards Sales and Customer Services, 389 Chiswick High Road, London W4 4AL
Tel: 0181 996 7000; fax: 0181 996 7001

BSI Standards Catalogue: annual list of British, international and European Standards and related publications, also describes BSI organization and services; lists updated monthly in *BSI News*. There is also on subscription BSI PLUS, an updating service tailored to a user's needs. With AFNOR and DIN, BSI is developing for selected areas a series of CD-ROM products. It also publishes a number of standards handbooks covering matters such as building, quality assurance, and reliability and maintainability.

RIBA Publications, 39 Moreland Street, London EC1V 8BB
Tel: 0171 251 0791; fax: 0171 608 2375

Architect's Reference Annual: incorporating the RIBA List of Recommended Books prepared by the British Architectural Library. Annual reference and booklists, including a round-up of new publications and a directory of the RIBA and associated companies.

Other sources

BBA Publications, including lists of Agréments issues, and Methods of Assessment and Testing. BBA, PO Box 195, Bucknalls Lane, Garston, Watford WD2 7NG; tel: 01923 670844; fax: 01923 662133.
Building Bookshop Catalogue: published annually. The Building Centre, 26 Store Street, London WC1E 7BT.
CIBSE Publications Catalogue: lists ASHRAE, BSRIA and HEVAC as well as CIBSE publications. CIBSE Bookshop, 222 Balham High Road, London SW12 9BS.
CIOB Mail Order Catalogue: CIOB Bookshop, Englemere, Kings Ride, Ascot SL5 8BJ; tel: 01344 874640; fax: 01344 23467.
CIRIA Publications: 6 Storey's Gate, London SW1P 3AU.
E & F N Spon Architecture, Building and Surveying Catalogue: Chapman & Hall, 2–6 Boundary Row, London SE1 8HN.
HMSO Books: quarterly list of selected recent publications. HMSO Publications Centre, PO Box 276, London SW8 5DT.
HSE Health & Safety Executive booklist: HSE Information Centre, Broad Lane, Sheffield S3 7HQ
NHBC Standards and other publications: NHBC Distribution Centre, Buildmark House, Chiltern Avenue, Amersham, Bucks HP6 5AP.

Information series, often on subscription

BRE Digests: essential reference material for building industry. Individually priced or on subscription.
BRE Defect Action Sheets: description of common faults in house-building, and avoidance measures; replaced by *Good Building Guides*. Both series individually priced.
BRE Information Papers: summaries of recent research results and their application. Individually priced.
Building Research Bookshop, Garston, Watford WD2 7JR.

Architectural Periodicals Index: compiled by British Architectural Library; published quarterly by subscription.
RIBA Publications, 39 Moreland Street, London EC1V 6BB
BSI News: published monthly, includes in *BSI Update* lists of new and revised British Standards, European and international publications, amendments to *Buyers Guide* (*BSI in Europe*: periodically published with *BSI News*).
BSI Standards Sales and Customer Services, 389 Chiswick High Road, London W4 4AL.
Building Technical Files: discontinued series published by Building (Publishers) Limited, 1 Millharbour, London E14 9RA.
DoE Construction Monitor, incorporating *Euronews Construction*: periodical newsletter for UK construction industry and professions, issued free by Construction Policy Directorate, Department of the Environment, 2 Marsham Street, London SW1P 3EB. Tel: 0171 276 6688.
DTI The Single Market: series of occasional free booklets giving information on EU Directives etc. and addressed to UK industry including construction.
Department of Trade and Industry Single Market Unit, 123 Victoria Street, London SW1E 6RB; DTI 1992, PO Box 1992, Cirencester GL7 1RN; or by telephone from 0181 200 1992.
RIBA Practice: periodical newsletter on legislation, standards and other professional matters, distributed to members as supplement to *RIBA Journal*. Practice Department, RIBA, 66 Portland Place, London W1N 4AD.

Advisory and information Services, usually on subscription

BSRIA EuroCentre: a service of publications, market studies etc. on legislation and standards affecting the building services industry. BSRIA, Old Bracknell Lane West, Bracknell RG12 4AH.
National Building Specification: a complete specification writing system available on subscription from NBS Services Ltd, Mansion House Chambers, The Close, Newcastle upon Tyne NE1 3RE.
RIBA Office Library Service: an installed and maintained product information library, available on subscription from 4 Park Circus Place, Glasgow G3 6AY.
RIBA Technical Information Microfile: a microfiche library of British Standards and Codes, BRE Digests and other technical material, updated three times a year and available on subscription. RIBA Publications, 39 Moreland Street, London EC1V 8BB.
Technical Indexes Ltd, Willoughby Road, Bracknell RG12 8DW (tel: 01344 426311; fax: 01344 424971).

Index

Page numbers in **bold** refer to figures, page numbers in *italic* refer to tables and page numbers in square brackets ([ ]) refer to boxes.